T0413351

FOOD SAFETY

Rapid Detection and Effective Prevention of Foodborne Hazards

FOOD SAFETY

Rapid Detection and Effective Prevention of Foodborne Hazards

Edited by

Lan Hu, MD, PhD

Apple Academic Press Inc.
3333 Mistwell Crescent
Oakville, ON L6L 0A2 Canada

Apple Academic Press Inc.
9 Spinnaker Way
Waretown, NJ 08758 USA

Exclusive worldwide distribution by CRC Press, a member of Taylor & Francis Group

International Standard Book Number-13: 978-1-77188-628-4 (Hardcover)
International Standard Book Number-13: 978-1-315-10909-1 (eBook)

Library and Archives Canada Cataloguing in Publication

Food safety (Oakville, Ont.)
Food safety : rapid detection and effective prevention of foodborne hazards / edited by Lan Hu, MD, PhD.
Includes bibliographical references and index.
Issued in print and electronic formats.
ISBN 978-1-77188-628-4 (hardcover).--ISBN 978-1-315-10909-1 (PDF)
1. Foodborne diseases--Prevention. 2. Food--Microbiology. I. Hu, Lan (Microbiologist), editor II. Title.
QR201.F62F66 2018 615.9'54 C2018-900493-2 C2018-900494-0

CIP data on file with US Library of Congress

Apple Academic Press also publishes its books in a variety of electronic formats. Some content that appears in print may not be available in electronic format. For information about Apple Academic Press products, visit our website at **www.appleacademicpress.com** and the CRC Press website at **www.crcpress.com**

ABOUT THE EDITOR

Lan Hu, MD, PhD

Lan Hu, MD, PhD, is a microbiologist and medical officer with over 20 years of laboratory and research experience in bacteriology, molecular biology, infectious disease, and food safety. She has extensive knowledge in multiple fields, including medicine, microbiology, biology, immunology, and epidemiology. Trained formally on how the U.S. Food and Drug Administration (FDA) reviews biology, drugs, food, and related devices, she has in-depth knowledge of the regulations, laws, and procedures related to products and public health. Dr. Hu has worked at several research institutes such as the National Institutes of Health (NIH), the Naval Medical Research Center (NMRC), and the FDA. She has initiated, directed, and completed numerous research projects and programs, particularly associated with developing pathogen detection methods and investigating bacterial pathogen–host cell interaction and pathogenic mechanisms.

Dr. Hu has received several awards for her work, including a Fellows Award for Research Excellence (1999 and 2000) from the NIH and the Young Investigator Award from the 10th International Workshop on *Campylobacter*, *Helicobacter*, and Related Organisms. Dr. Hu has written a number of articles and book chapters and also acts as a reviewer for several scientific journals, including *Infection and Immunity* and *Food Protection.*

CONTENTS

List of Contributors .... *ix*

List of Abbreviations .... *xi*

Preface .... *xv*

PART I: BIOLOGICAL TOXINS **1**

1. Staphylococcal Enterotoxins: Food Poisoning and Detection Methods **3**

Xin Wang and Yinduo Ji

PART II: FOODBORNE BACTERIAL PATHOGENS **23**

2. *Salmonella* Species **25**

Lan Hu and Baoguang Li

3. *Campylobacter* Species **55**

Lan Hu and Dennis D. Kopecko

4. *Escherichia coli* O157:H7 .... **93**

Xiangning Bai and Yanwen Xiong

5. Diarrhea and Enterotoxigenic *Escherichia coli* **123**

Xin-He Lai, Long-Fei Zhao, and Yan Qian

6. *Listeria monocytogenes* as Foodborne Pathogen: Genetic Approaches, Identification, and Detection Methods **151**

Hossam Abdelhamed, Seongwon Nho, Attila Karsi, and Mark L. Lawrence

7. *Shigella*: A Threat to the Food Industry **179**

Sushma Gurumayum, Sushree Sangita Senapati, Prasad Rasane, and Sawinder Kaur

PART III: FOODBORNE VIRUSES **203**

8. Norovirus **205**

Yuan Hu and Haifeng Chen

Index **237**

LIST OF CONTRIBUTORS

Hossam Abdelhamed
Department of Basic Sciences, College of Veterinary Medicine, Mississippi State University, Starkville 39762, MS, USA

Xiangning Bai
Researcher, Collaborative Innovation Center for Diagnosis and Treatment of Infectious Disease, State Key Laboratory of Infectious Disease Prevention and Control, National Institute for Communicable Disease Control and Prevention, Chinese Center for Disease Control and Prevention, Beijing, PR China

Haifeng Chen
Microbiologist, Division of Molecular Biology, Center for Food Safety and Applied Nutrition, Food and Drug Administration, Laurel 20708, MD, USA

Sushma Gurumayum
Department of Microbiology, College of Allied Health Sciences, Assam Down Town University, Panikhaiti, Guwahati 781026, India

Lan Hu
Research Microbiologist, Division of Molecular Microbiology, CFSAN, FDA, Laurel, MD 20708, USA

Yuan Hu
Research Microbiologist, Northeast Region Laboratory Office of Regulatory Affairs, Food and Drug Administration, Jamaica, NY, USA

Yinduo Ji
Professor, Department of Veterinary Biomedical Sciences, College of Veterinary Medicine, University of Minnesota St. Paul 55108, MN, USA

Attila Karsi
Department of Basic Sciences, College of Veterinary Medicine Mississippi State University, Starkville 39762, MS, USA

Sawinder Kaur
Department of Food Technology and Nutrition, Lovely Professional University, Jalandhar, Punjab 144411, India

Dennis J. Kopceko
President, CombiVax, LLC, 16325 Whitehaven Rd, Silver Spring 20906, MD, USA

Xin-He Lai
Scientist, Department of Pediatrics, the First Affiliated Hospital of Wenzhou Medical University, Wenzhou, PR China

Mark L. Lawrence
Professor, Department of Basic Sciences, College of Veterinary Medicine, Mississippi State University, Starkville 39762, MS, USA

Baoguang Li
Research Microbiologist, Division of Molecular Microbiology, OARSA, CFSAN, FDA, Laurel 20708, MD, USA

Seongwon Nho
Department of Basic Sciences, College of Veterinary Medicine, Mississippi State University, Starkville 39762, MS, USA

Yan Qian
Department of Pediatrics, The First Affiliated Hospital of Wenzhou Medical University, New Campus at Nanbaixiang, Ouhai District, Wenzhou, China 325000

Prasad Rasane
Assistant Professor, Centre of Food Science and Technology, Banaras Hindu University, Varanasi 221005, Uttar Pradesh, India.

Sushree Sangita Senapati
Department of Microbiology, College of Allied Health Sciences, Assam Down Town University, Panikhaiti, Guwahati 781026, India

Xin Wang
College of Food Science and Engineering, Northwest A&F University, Yangling 712100, Shaanxi, PR China

Yanwen Xiong
Senior researcher, Collaborative Innovation Center for Diagnosis and Treatment of Infectious Diseases, State Key Laboratory of Infectious Disease Prevention and Control, National Institute for Communicable Disease Control and Prevention, Chinese Center for Disease Control and Prevention, Beijing, PR China

Long-Fei Zhao
College of Life Sciences, Key Laboratory of Plant-Microbe Interactions of Henan, Shangqiu Normal University, 55 Pingyuanzhong Road, Shangqiu, Henan, PR China 476000

LIST OF ABBREVIATIONS

A/E lesions	attaching and effacing lesions
APCs	antigen presenting cells
Aw	water activity
BAM	bacteriological and analytical method
CAMP	Christie–Atkins–Munch–Peterson
CDT	cytolethal distending toxin
CDC	Centers for Disease Control and Prevention
CCDA agar	charcoal cefoperazone deoxycholate agar
CFA	colonization factor
CFTR	cystic fibrosis transmembrane conductance regulator
CFs	colonization factors
CFUs	colony forming units
CPS	capsular polysaccharide
CRP	cAMP receptor protein
cs	centisomes
CSTE	Council of State and Territorial Epidemiologists
C3b	complement component 3 fragment b
CT	cholera toxin
DAEC	diffusely adherent *E. coli*
DALYs	disability-adjusted life year
DCs	dendritic cells
DEC	diarrheagenic *E. coli*
EAEC	enteroaggregative *E. coli*
ECB	*E. coli* broth
EHEC	enterohemorrhagic *E. coli*
EIEC	enteroinvasive *E. coli*
ELFA	enzyme-linked fluorescent assay
ELISA	enzyme-linked immunosorbent assays
EM	electron microscopy
EPEC	enteropathogenic *E. coli*
EPA	Environmental Protection Agency
ERS	Economic Research Services
ESI	electrospray ionization

eta and *etb*	exfoliative toxin A and B
ETEC	enterotoxigenic *E. coli*
FAB	fast atom bombardment
FCV	feline calicivirus
FDA	United States Food and Drug Administration
5-HT	5-hydroxytryptamine
GBS	Guillain–Barré syndrome
GC-C	guanylate cyclase C
GSP	general secretory pathway
HACCP	hazard analysis and critical control point
HBGAs	human histo-blood group antigens
HC	hemorrhagic colitis
Hcp	hemolysin co-regulated proteins
HE agar	Hektoen enteric agar
HPLC	high pressure liquid chromatography
HUS	hemolytic uremic syndrome
IBD	inflammatory bowel disease
IBS	post-infectious irritable bowel syndrome
IL	interleukin
IMS	immunomagnetic separation
iNTS	invasive non-typhoid *Salmonella*
IPEC-J2	porcine intestinal epithelial cells
IS	insertion sequences
ISO	International Organization for Standardization
KIA	Kligler iron agar
LAMP	loop-mediated isothermal amplification
LEE	locus of enterocyte effacement
LOS	lipooligosaccharides
LPS	lipopolysaccharide
LT enterotoxin	heat-labile enterotoxin
M cell	microfold cell
MALDI	matrix-assisted laser desorption/ionization
MALDI-TOF-MS	matrix-assisted laser desorption ionization-time of flight mass spectrometry
MeOPN	O-methyl phosphoramidate
MHC	major histocompatibility complex
MLST	multilocus sequence typing
MLVA	multiple-locus variable-number tandem repeat analysis

MNV	murine norovirus
MRSA	methicillin-resistant *S. aureus*
MPCR-DHPLC	multiplex polymerase chain reaction coupled with denaturing high performance liquid chromatograph
MS	mass spectrometry
NLRs	NOD-like receptors
nqPCR	nested quantitative PCR
NTS	nontyphoid *Salmonella*
OCLA	Oxoid Chromogenic *Listeria* Agar
OMVs	outer membrane vesicles
ORFs	open reading frames
ORT	oral rehydration treatment solution
PCR	polymerase chain reaction
PCR-ELISA	polymerase chain reaction-enzyme linked immuno-sorbent assay
PD	plasma desorption
PFGE	pulse-field gel electrophoresis
PI	isoelectric point
PGE_2	prostaglandin E2
PKA	protein kinase A
PKC	protein kinase C
PMNs	polymorphonuclear leukocytes
PMQR	plasmid-mediated quinolone resistance
PRRs	pattern recognition receptors
PT	phage typing
qPCR	quantitative polymerase chain reaction
qRT-PCR	quantitative real-time polymerase chain reaction
ReA	Reiter's syndrome
RPLA	reverse passive latex agglutination
RT-PCR	reverse transcription-polymerase chain reaction
SaPIs	*Staphylococcus aureus* pathogenicity islands
SscDNA	single-stranded norovirus cDNA
SCC	staphylococcal cassette chromosome
SCV	*Salmonella*-containing vacuole
SDS PAGE	sodium dodecyl sulfate polyacrylamide gel electrophoresis
SEA	staphylococcal enterotoxin A
SEs	staphylococcal enterotoxins
SFP	staphylococcal food poisoning

SFPOs	staphylococcal food poisoning outbreaks
SMAC	sorbitol MacConkey agar
SNP	single nucleotide polymorphism assays
SPIs	*Salmonella* pathogenicity islands
SPR	surface plasmon resonance
SS agar	*Salmonella–Shigella* agar
ST	sequencing type
ST enterotoxin	heat-stable enterotoxin
STEC	Shiga toxin-producing *Escherichia coli*
Stx	Shiga toxin
T5SS	type 5 secretion system
TSA	tryptone soy agar
TLRs	Toll-like receptors
TNF	tumor necrosis factor
TPS	two-partner secretion
TSB	tryptone soy broth
T3SS	type 3 secretion system
T6SS	type 6 secretion system
T2SS	type 2 secretion system
TSST-1	toxic shock syndrome toxin-1
USDA	United States Department of Agriculture
Vß	variable region
VBNC bacteria	viable but nonculturable bacteria
VgrG	valine-glycine repeat G
VLPs	virus-like particles
VRBG	violet red bile glucose agar
VSP	Vessel Sanitation Program
VTEC	verocytotoxin-producing *E. coli*
WGS	whole genome sequencing
WK	White–Kauffmann-antigenic scheme
XLD agar	xylose-lysine-deoxycholate agar
XLT-4 agar	xylose lysine tergitol 4 agar

PREFACE

Foodborne disease is a major public health and economic burden in developed and developing countries. In the United States, the incidence of foodborne disease is approximately 9.4 million cases with about 56,000 hospitalizations and 1351 deaths every year.[1] Globally about 582 million cases with 25.2 million DALYs (disability-adjusted life years) are caused by the consumption of contaminated food annually, most of which occur in Asia and Africa.[2,3]

In the United States, the Center for Disease Control and Prevention recognizes 31 major pathogens that caused the foodborne disease. These pathogens include norovirus (which causes 58% of cases), nontyphoid *Salmonella* (NTS) (11%), *Clostridium perfringens* (11%), and *Campylobacter* (9%), as well as enteropathogenic *Escherichia coli*, enterotoxigenic *E. coli*, and *Vibrio cholerae*.[1] The major causes of DALYs are NTS (28%), *Toxoplasma gondii* (24%), *Listeria monocytogenes* (19%), and norovirus (11%).[1] Foodborne illness is also attributable to chemical contaminants such as heavy metals (mercury, lead, arsenic), chemical compounds (melamine and its analogs), and biological toxins (staphylococcal enterotoxins, mycotoxin, botulinum neurotoxins, epsilon toxin).

Understanding the up-to-date epidemiologic distribution, pathogenesis, genomics, and detection methods of these hazards is necessary for preventing and managing foodborne pathogens. The aim of this book is to fill a gap in basic food safety knowledge and new applications of analytical and molecular biology techniques employed to detect and prevent chemical and microbial hazards in foods. We will provide an overview of the general concepts, mechanisms, and new applications of analytical and molecular biological techniques, focusing on the main biological hazards.

Today, food safety is intensely concerned not only by food-related professionals and policymakers but also by the public. More and more people are interested in this important issue that influences their health and quality of life. Foodborne hazards including chemical and biological hazards cause food intoxication, infectious diseases, cancers, and other diseases.

This book is designed: (1) to provide the most recent knowledge and techniques in foodborne hazard analyses, which are useful for professionals and other specialists who are working in academia, clinical laboratories, food safety and manufacturing sectors; (2) to give a useful guide for talented students from middle schools to graduate education who are interested in food safety research and who work on food safety research projects; and (3) to present up-to-date knowledge of effective detection, reduction, and prevention strategies in managing foodborne hazards.

This book is written in an easy-to-read style. The first part of each chapter introduces one of foodborne hazards such as: What kind of diseases does the foodborne hazard cause? Why is it necessary for us to study it? What routes does it enter our food and how does it cause disease? The latter portion of each chapter addresses the current applications of new technologies to isolate, detect, and identify the hazard, and the preventive procedures. For example, how can the current techniques be used to detect the foodborne hazard? How do we prevent diseases caused by the foodborne hazard? Lastly, food safety experts may also present their experimental results in detail.

The features of the book are: (1) each book chapter covers thoroughly one foodborne hazard from basic knowledge to future direction; (2) it covers the latest rapid, sensitive, specific, and reliable methods and techniques for detection, analysis, and identification of the various foodborne hazards from food and drink; (3) experts will share their precious insights and scientific data in detecting, reducing, and preventing foodborne hazards; (4) the book targets multiple readers; (5) it is user-friendly educational material of foodborne hazards for training purpose.

As the editor of this book, I greatly appreciate Apple Academic Press for giving me the precious opportunity to work on such an important topic, which influences public health and everyone's quality of life. I would also like to thank all authors and reviewers who contribute their knowledge and experience to the book. I hope the book would be favored by all kinds of readers.

—**Lan Hu,** MD, PhD
CFSAN, FDA, Laurel, MD

REFERENCES

1. Scallan, E.; Hoekstra, R. M.; Angulo, F. J.; Tauxe, R. V.; Widdowson, M. A.; Roy, S. L.; Jones, J. L.; Griffin, P. M. Foodborne Illness Acquired in the United States—Major Pathogens. *Emerg. Infect. Dis.* **2011,** *17,* 7–15.

2. Crump J. A.; Luby S. P.; Mintz E. D. The Global Burden of Typhoid Fever. *Bull. World Health Organ.* **2004,** *82,* 346–353.
3. Kirk, M. D.; Pires, S. M.; Black, R. E.; Caipo, M.; Crump, J. A.; Devleesschauwer, B.; Döpfer, D.; Fazil, A.; Fischer-Walker, C. L.; Hald, T.; Hall, A. J.; Keddy, K. H.; Lake, R. J.; Lanata, C. F.; Torgerson, P. R.; Havelaar, A. H.; Angulo, F. J. World Health Organization Estimates of the Global and Regional Disease Burden of 22 Foodborne Bacterial, Protozoal, and Viral Diseases, 2010: A Data Synthesis. *PLoS Med.* **2015,** *12* (12), e1001921.

PART I
Biological Toxins

CHAPTER 1

STAPHYLOCOCCAL ENTEROTOXINS: FOOD POISONING AND DETECTION METHODS

XIN WANG[1] and YINDUO JI[*2]

[1]*College of Food Science and Engineering, Northwest A&F University, Yangling 712100, Shaanxi, PR China*

[2]*Department of Veterinary Biomedical Sciences, College of Veterinary Medicine, University of Minnesota, St. Paul 55108, MN, USA*

[*]*Corresponding author. E-mail: jixxx002@umn.edu*

CONTENTS

Abstract 4
1.1 Introduction 4
1.2 Staphylococcal Food Poisoning (SFP) 5
1.3 Staphylococcal Enterotoxins and Their Characteristics 6
1.4 Animal Models and Cell-Based Approaches for Examination of Staphylococcal Enterotoxins 10
1.5 Molecular Methods 11
1.6 Immunological Methods 14
1.7 Mass Spectrometry-Based Methods 16
1.8 Conclusions 16
Keywords 17
References 18

ABSTRACT

Staphylococcus aureus can cause a variety of diseases including skin and soft tissue infections, and systematic and life-threatening infections. Meanwhile, *S. aureus* is one of the major foodborne pathogens causing food poisoning both in humans and animals. The pathogenesis of *S. aureus* causing food poisoning is attributable to the production of many staphylococcal enterotoxins (SEs), SEs-like, exfoliative toxin A and B and/or toxic shock syndrome toxin-1 by some *S. aureus* isolates. Therefore, the development of rapid, specific, sensitive, and reliable diagnostic approaches is not only important for detecting and distinguishing staphylococcal exterotoxins, but also can provide powerful tool for food safety control in the public health and in food industry. In this chapter, we discuss the staphylococcal food poisoning, predominant genotypes of *S. aureus* isolates, SEs and their characteristics, animal models, and various cell-based strategies to evaluate SEs. Moreover, we review molecular-based methods, immunological approaches, as well as advanced mass spectrometry technologies that have been utilized in the determination and detection of SEs for food safety.

1.1 INTRODUCTION

Staphylococcus aureus is a major pathogen that causes a variety of diseases including skin and soft tissue infections, and systematic and life-threatening infections, such as pneumonia, endocarditis, and toxic shock syndrome. This organism has caused tens of millions of infections annually in the United States. The continuing emergence of multiple drug resistant *S. aureus* isolates, especially methicillin resistant *S. aureus* (MRSA), has led to a serious public concern. MRSA accounts for 40–60% of the hospital-acquired infections in the United States and has increasingly spread beyond healthcare facilities, emerging as a community-acquired pathogen that results in minor skin and soft tissue infections as well as severe invasive diseases, such as necrotizing pneumonia and a sepsis syndrome. The pathogenicity of *S. aureus* partially relies on the coordinately regulated expression of virulence factors that allows the bacterium to evade the host immune system and/or promote survival during infection. Similar to other bacterial pathogens, *S. aureus* has evolved a series of regulatory effectors that allow the organisms to sense and adapt to varying environmental stimuli and survive within a particular niche by modulating specific cellular responses and virulence gene expression.

Meanwhile, *S. aureus* is one of the major foodborne pathogens causing food poisoning both in humans and animals.[1] It has been well established that the pathogenesis of *S. aureus* causing food poisoning is because some *S. aureus* isolates produce many staphylococcal enterotoxins (SEs), staphylococcal enterotoxin-like (SEls), exfoliative toxin A and B (*eta* and *etb*) and toxic shock syndrome toxin-1 (TSST-1). SEs or SEls are well known as a major cause of food poisoning.[2] At least 23 different types of SEs have been identified up to today, including the classical SEs (SEA through SEE), and the new types of SEs (SEG to SEX).[2]

S. aureus often colonizes on the skin and mucous membranes of humans as commensal inhabitants. Thus, the food workers carrying enterotoxin-producing *S. aureus* in their noses and/or hands are the major resource of food contamination due to poor hygiene during the food preparation and process.[2] SEs are also a threat to both food safety and food security if they are produced in a purified form that can be used as a deliberate adulterant. Therefore, it is necessary to develop reliable, sensitive, and rapid methods for the detection of SEs. A large number of sensitive and selective detection methods based on different principles have been reported. This review provides a brief overview of conventional methods and focuses on immunosensors, which are currently used to detect SEs in food. Finally, future trends and conclusions are discussed.

1.2 STAPHYLOCOCCAL FOOD POISONING (SFP)

SFP is caused by the ingestion of SEs produced during massive growth of *S. aureus* in food. The severity of SFP mainly depends on individual's sensitivity to the toxins, the amount of contaminated food eaten, the amount of toxin in the food ingested, and the general health conditions. The incubation period of illness ranges from 30 min to 8 h, but usually 2–4 h.[3] SFP induces a rapid onset of symptoms including nausea, vomiting, retching, abdominal cramping, and prostration, which are often accompanied by diarrhea and sometimes fever. In severe cases, patients may have headache, muscle cramping, severe fluid, and electrolytes loss with weakness and low blood pressure or shock. SFP is usually self-limiting within two days, but occasionally becomes severe enough to warrant hospitalization for a longer period of time.[4]

S. aureus frequently colonizes on the skin and mucous membranes of humans with approximately 25–30% for persistent colonization.[4,5] Thus, food handlers carrying enterotoxin-producing *S. aureus* in their nasal cavity or on their skin are a recognized risk factor for food contamination during food preparation, processing and handling, followed by storage under conditions which allow *S. aureus* growth and production of the enterotoxin(s).[6]

Various foods have been identified to be associated with SFP, including meat and meat products; poultry and egg products; salads; cream-filled pastries; sandwich fillings; and milk and dairy products,[7] Chinese frozen dumplings.[8] However, the foods most often involved in outbreaks differ widely from one country to another because of the variation in food consumption culture and habits. For example, in England and Wales, 60% of the staphylococcal food poisoning outbreaks (SFPOs) between 1992 and 2009 were reported due to consuming of contaminated poultry meat and red meat;[3] whereas in Japan, 36% of the SFPOs between 1995 and 1999 resulted from contaminated grains like rice balls and composite ready-to-eat food, 5.6% of the incidents were due to contaminated fish and shellfish, and less than 1% resulted from contaminated milk products.[3]

More importantly, it has been discovered that *S. aureus* isolates from SFPOs of different countries exhibit different genetic features or genotypes, such as sequencing type (ST). For example, ST1, ST81, and ST6 *S. aureus* isolates are prevalent epidemic clones induced SFPOs in South Korea,[9] Japan,[10,11] and China,[12,13] respectively. One possible explanation is that the different genotypes prevalence of *S. aureus* SFPOs is likely due to the differences in food consumption and eating habits in these countries.[9] However, the predominant *S. aureus* isolates in SFPOs are not necessarily the predominant *S. aureus* lineage that causes illness in hospitals and animals in these countries.[13] The reason why these ST types became the dominant genotypes of *S. aureus* isolates that causes SFP is still undefined. It is still unclear whether these ST types are the predominant genotypes of *S. aureus* isolates in food products and/or among SFP isolates in different counties.[13]

1.3 STAPHYLOCOCCAL ENTEROTOXINS AND THEIR CHARACTERISTICS

SEs are potent gastrointestinal exotoxins, which are expressed by *S. aureus* throughout the logarithmic phase of growth or during the transition from the exponential to the stationary phase. To date, 23 distinct enterotoxins have been identified based on their antigenicity and they have sequentially been assigned a letter of the alphabet in order of their discovery (SEA to SEX): enterotoxins A (SEA), B (SEB), C (SEC1, SEC2, and SEC3), D (SED), E (SEE), G (SEG), H (SEH), I (SEI), J (SElJ), K (SElK), L (SElL), M (SElM), N (SElN), O (SElO), P (SElP), Q (SElQ), R (SER), S (SES), T (SET), U (SElU), and U2 (SEW), V (SElV), and X (SElX).[14] They belong to a group of single-chain proteins with low molecular weights ranging from 25 to 29 kDa, sharing common phylogenetic relationships, structure, function, and sequence homologue.[15] STs possess two important biological activities, including that they cause gastroenteritis in the gastrointestinal tract, as well as function as superantigens in the immune system. Functional enterotoxins are able to bind to the alpha-helical regions of the major histocompatibility complex (MHC) class II molecules outside the peptide-binding groove of the antigen presenting cells (APCs), and also bind to the variable region (Vß) on T-cell receptors, resulting in CD^{4+} T-cell activation, immune modulation, and consequently toxic shock syndrome.[16,17]

SEs are soluble in water and salt solution. The pH values of SEs range from 7.0 to 8.6. SEs are highly stable, resist to most proteolytic enzymes such as pepsin and trypsin, so that they can maintain their activity in the digestive tract after ingestion. Moreover, SEs are highly heat resistant, as they still possess the biological activities after boiling at 100°C for at least 30 min and probably longer. The biological activity of SEs could be abolished after heating at 218–248°C for 30 min. Although pasteurization and cooking kills staphylococci cells, SEs in the food generally retain their biological activity as they are thermo-stable. Thus, SFP cases might occur when no viable bacteria are isolated from the cultures of suspected foodstuff. Therefore, it is not surprising when obtaining a negative staphylococci culture result, but a positive SEs detection during examination of food product.[16]

All SEs and SEls genes of *S. aureus* are located at different accessory genetic elements, including plasmids, prophages, pathogenicity islands (SaPIs), genomic island vSa, or next to the staphylococcal cassette chromosome (SCC) elements (Table 1.1). Most of these are mobile genetic elements, and their transferring among *S. aureus* isolates can increase the capacity of *S. aureus* to resist antibiotics, evade host immune system and cause disease, as well as contribute to the evolution of this important

TABLE 1.1 Characteristics and Gene Size and Encoding Gene Location of SEs and SEIs.[2,18–21]

SE gene	Strain	GenBank accession no.	Molecular mass (kDa)	Gene length (bp)	Emetic activity	Location of gene	Accessory genetic element
sea	MW2	NC-003923	27.1*	774	Yes	Genome	ΦSa3ms, ΦSa3mw, Φ252B, ΦNM3, ΦMu50a
seb	Col	NC-002951	28.4	801	Yes	Genome/ Plasmid	pZA10, SaPI3
sec	Mu3	NC-009782	27.5–27.6*	801	Yes	Genome	SaPIn1, SaPIm1, SaPImw2, SaPIbov1
sed	Plasmid pIB485	M94872.1	26.9	777	Yes	Plasmid	pIB485-like
see	ATCC27664	AY518387.1	26.4	774	Yes	Genome	ΦSa[b]
seg	Mu3	NC-009782	27.0*	777	Yes	Genome	egc1 (vSaβ I); egc2 (vSaβ III); egc3; egc4
seh	MSSA474	NC-002953	25.1	726	Yes	Genome	MGEmw2/mssa476 seh/Δseo
sei	Mu3	NC-009782	24.9	777	Weak	Genome	egc1 (vSaβ I); egc2 (vSaβ III)); egc3
selj	Plasmid pF5	AB330135.1	28.5	807	nd	Plasmid	pIB485-like; pF5
selk	MRSA252	BX571865	26.0*	774	nd	Genome	ΦSa3ms, ΦSa3mw, SaPI1, SaPI3, SaPIbov1, SaPI5
sell	Mu3	NC-009782	26.0	723	No[a]	Genome	SaPIn1, SaPIm1, SaPImw2, SaPIbov1
selm	Mu3	NC-009782	24.8	720	nd	Genome	egc1 (vSaβ I); egc2 (vSaβ III)
seln	Mu3	NC-009782	26.1	777	nd	Genome	egc1 (vSaβ I); egc2 (vSaβ III); egc3; egc4
selo	Mu3	NC-009782	26.7	783	nd	Genome	egc1(vSaβI);egc2(vSaβIII);egc3;egc4;M GEmw2/mssa476seh/Δseo
selp	N315	NC-002745	27.0	783	nd[a]	Genome	ΦN315, ΦMu3A
selq	USA300-FPR3757	NC-007793	25.0	729	No	Genome	ΦSa3ms, ΦSa3mw, SaPI1, SaPI3, SaPI5

TABLE 1.1 *(Continued)*

SE gene	Strain	GenBank accession no.	Molecular mass (kDa)	Gene length (bp)	Emetic activity	Location of gene	Accessory genetic element
ser	Plasmid pF5	AN330135.1	27.0	780	Yes	Plasmid	pIB485-like; pF5; pF6; pF73
ses	Plasmid pF5	AB330135.1	26.2	774	Yes	Plasmid	pF5
set	Plasmid pF5	AB330135.1	22.6	651	Weak	Plasmid	pF5
selu	352E	AY205305.1	27. 1*	780	nd	Genome	egc2 (vSaβ III); egc3
selu2(sew)	A900624	EF030428.1	nd	771	nd	Genome	egc4
selv	A900624	EF030427.1	nd	720	nd	Genome	egc4
selx	USA30-FPR3757	HQ850968.1	19.3	612	nd	Genome	Core genome

Note: nd—not determined; the symbol "*" represents variants.

pathogen.[17] For example, *sed* gene is only located at the plasmids, whereas *seb*, *selj*, *ser*, *ses*, and *set* genes have been found in the chromosome, but could also be plasmid borne. Other SEs and SEIs genes are located at genome.

1.4 ANIMAL MODELS AND CELL-BASED APPROACHES FOR EXAMINATION OF STAPHYLOCOCCAL ENTEROTOXINS

1.4.1 ANIMAL MODELS

In order to establish a cheap and specific test method of SEs in vivo, a variety of animal models have been utilized, including pigs, dogs, cats, kittens, and monkeys. It has been reported that all of these animals, with the exception of monkey-feeding and kitten-intraperitoneal tests,[22,23] were relatively insensitive to the enterotoxins. Monkey-feeding and kitten-intraperitoneal tests have been widely used in public health and in the pathogenicity examination of the nature of enterotoxins. However, these animal models showed low sensitivity and specificity. A sensitization test of SEB was developed by using the animal models of house musk shrews (*Suncus murinus*)[20,26] and guinea pigs,[24] respectively. Unfortunately, the other SEs failed to produce a similar sensitization response to that of SEB. Thus, it is not practical to use the above method to evaluate and examine SEs. Overall, the use of animal tests is not appropriate for characterizing SFPOs because of poor sensitivity, specificity, and the expensive costs of the animals.

Tollersrud et al. reported that two cows infected with *S. aureus* strain M60 shed superantigen enterotoxin D in their milk. Moreover, it was found that the animals that secreted enterotoxin D in their milk induced the highest enterotoxin D serum antibody.[25] It was reported that superantigen enterotoxin D was able to cause a mitogenic effect on bovine leukocytes, while adding anti-enterotoxin D antibodies could compromise its impact on leukocytes.[25] Thus, it is necessary to investigate the possibility of vaccines based on SEs in humans because of their association with food poisoning and toxic shock syndrome.[26]

The emetic responses induced by staphylococcal enterotoxin A (SEA), SEB, SEC2, SED, SEE, SEG, SEH, SEI SER, SES, SET, and SEY were tested using the house musk shrew (*Suncus murinus*). SEA, SEE, and SEI toxins showed higher emetic activity than the other SEs. SEB, SEC2, SED,

SEG, SEH, SER, SES, SET, and SEY also induced emetic responses in this animal model, but relatively high doses were required (Table 1.2).[18,27,28] In addition, the emetic responses induced by staphylococcal enterotoxin A (SEA), SEB, SEK, SEL, SEM, SEN, SEO SEP, SEQ, SER, SES, and SET were examined using the monkey model. SEA and SEB showed higher emetic activity than the other SEs, while relatively high doses of SEK, SEL, SEM, SEN, SEO, SEP, SEQ, SER, SES, and SET were necessary to cause emetic responses in this animal model.

1.4.2 CELL-BASED ASSAY

Despite the fact that many types of animals such as shrews, monkey, kitten, and guinea pigs were used to test the biological activity of SEs, high cost of animals, low sensitivity and poor reproducibility make it impractical to examine a large number of samples, as well as raise ethical concerns with regard to the use of experimental animals.[32] Moreover, other approaches, such as immunological methods and MS, cannot distinguish with active and inactive toxins, and result in false-positive results with food samples.[33] Thus, other alternative strategies, including in vivo monkey and kitten bioassays, have been developed for the detection of the biological activity of SEs.[32,34–36] Among these, the cell-based assays has emerged as a rapid sensitive and quantitative bioassays for the detection of the biological activity of SEs, such as spleen T-cell proliferation bioassay,[34] lymphocyte T-cell proliferation.[36]

1.5 MOLECULAR METHODS

Molecular biological methods often involve nucleic acid hybridization and polymerase chain reaction (PCR) techniques. Gene specific oligonucleotide probes were first used to detect genes encoding enterotoxins in *S. aureus* strains isolated from contaminated foods and differentiate *S. aureus* isolates based upon nucleic acid hybridization approach.[37] Since then, dot blot hybridization techniques have been developed and applied to determine SE genes in *S. aureus* strains. However, the DNA hybridization strategy has an apparent drawback as this approach involve too many steps and takes a long time to obtain a final result. It is not practical and now is not utilized for the identification of SE genes in *S. aureus* strains.

TABLE 1.2 Emetic Activities of SEs on House Musk Shrews[18,27,28], Monkeys[18,29], Kittens[30], and Guinea Pigs.[24,31]

SEs	No. of shrewsa		SEs	No. of monkeysa		SEs	No. of kittensb		SEs	No. of guinea pig skin site	
	Dose (μg/ animal)	**Vomit / tested**		**Dose (μg/ kg)**	**Vomit/ tested**		**Dose (μg/ kg)**	**Vomit/ tested**		**Dose (ng/ skin site)**	**Diam. of skin reaction (mm)**
SEA	1	5/5	SEA	100	6/7	SEA	10	1/3	SEB	100	23, 25, 23, 20
	0.5	3/5		10	5/10		25	3/4		10	20, 19, 20, 15,
	0.3	1/6								1	
	0.1	0/6								0.1	17, 15, 17, 14, 9,
SEB	1000	3/3	SEB	100	4/4	SEB	10	0/5		0.01	13, 11, 11, 5
	200	2/5					25	2/5		0.0001	11, 9, 7, 0
	40	3/6					50	1/5			9, 5, 0, 0
	10	1/6					100	2/4			
	1	0/6									
SEC2	1000	2/2	SEK	100	2/6						
	200	0/3									
	40	0/4									
SED	1000	2/3	SEL	100	1/6						
	200	1/2									
	40	1/2									
	10	0/3									
SEE	10	2/2	SEM	100	1/7						
	1	0/2									
	0.1	0/2									

TABLE 1.2 *(Continued)*

SEs	No. of shrewsa		SEs	No. of monkeysa		SEs	No. of kittensb		SEs	No. of guinea pig skin site	
	Dose (μg/ animal)	Vomit / tested		Dose (μg/ kg)	Vomit/ tested		Dose (μg/ kg)	Vomit/ tested		Dose (ng/ skin site)	Diam. of skin reaction (mm)
SEG	1000	1/3	SEN	100	2/6						
	200	1/5									
	40	0/5									
SEH	1000	2/2	SEO	100	1/8						
	200	0/2									
	40	0/3									
SEI	10	5/5	SEP	100	3/6						
	1	2/6									
	0.1	0/6									
SER	1000	2/5	SEQ	100	2/6						
	200	0/5									
SES	100	2/3	SER	100	2/6						
	200	1/3									
	4	0/3									
SET	1000	1/5	SES	100	2/4						
	500	0/3									
SEY	500	4/7	SET	100	0/4						
	100	0/4									

[a]Intraperitoneal injection.

[b]Intravenous injections.

Different PCR-based approaches have been widely and successfully utilized for the detection of *S. aureus* including clinical methicillin-resistant *S. aureus* (MRSA).[38] PCR has the advantages of rapidity, high sensibility, high throughput, and good reproducibility. Thus, PCR-based technologies have been employed to monitor the food contamination by *S. aureus*, to determine, and trace food poisoning caused by SEs. Currently, several PCR variants have been developed to detect SEs genes, such as multiplex PCR,[29–32] real-time PCR,[39–41] reverse transcription real-time PCR (RT-PCR),[42] and loop-mediated isothermal amplification (LAMP),[43,44] combined multiplex loop-mediated isothermal amplification.[45] PCR also has the advantage of being combined with other techniques, such as PCR-enzyme linked immunosorbent assay (PCR-ELISA)[46] and multiplex PCR coupled with denaturing high performance liquid chromatograph (MPCR-DHPLC).[47] The detail information, including toxin genes, primers for PCR and product size is shown in Table 1.3. Using PCR-ELISA technologies we are not only able to determine the presence or absence of genes encoding SEs, but also can examine the expression levels of these SE genes in food samples. Moreover, PCR-ELISA technologies can be used to confirm *S. aureus* as causative agent in an outbreak.

1.6 IMMUNOLOGICAL METHODS

Immunological methods have been widely used to detect SEs produced by *S. aureus*, including enzyme-linked immunosorbent assay (ELISA),[56] enzyme-linked fluorescent assay (ELFA),[57] reverse passive latex agglutination (RPLA),[58] chemiluminescence immunoassays, electrochemiluminescence immunoassays,[59] surface plasmon resonance (SPR) immunoassays,[45,46] electrochemical immunoassay,[60] and immunomagnetic separation (IMS) method, multiplex electrochemiluminescence immunoassay.[48] These methods are used in the antigen–antibody reaction of SEs, which can effectively avoid the interference of impurities in the complex sample, simplifying the purification steps before processing, simple operation, high sensitivity and specificity, and other characteristics. To date, only antibodies against SEA to SEE, SEG, SEH,[61–63] SEII[64] and SElQ are available.[65] However, SEA through SEE can only be detected using commercial kits utilizing ELISA, ELFA, and RPLA. Therefore, immunological methods will not detect the other SEs, which could partly explain why some outbreaks remained uncharacterized without a known

TABLE 1.3 Oligonucleotide Primers of SEs Used in PCRs.

Gene	Forward primer sequence (5'-3')	Reverse primer sequence (5'-3')	Size (bp)	Reference
sea	AAAGATTTGCGAAAAAAGTGTGAATT	TAAATCGTAATTAACCGAAGGTTCTG	669	[48]
seb	GTATGGTGGTGTAACTGAGC	CCAAATAGTGACGAGTTAGG	164	[49]
sec	AGATGAAGTAGTTGATGTGTATGG	CACACTTTTAGAATCAACCG	451	[49]
sed	CCAATAATAGGAGAAAATAAAAG	ATTGGTATTTTTTTTCGTTC	278	[49]
see	AGGTTTTTTCACAGGTCATCC	CTTTTTTTTCTTCGGTCAATC	209	[49]
seg	TGCTATCGACACACTACAACC	CCAGATTCAAATGCAGAACC	704	[49]
seh	CGAAAGCAGAAGATTTACACG	GACCTTTACTTATTTCGCTGTC	495	[49]
sei	GACAACAAAACTGTCGAAACTG	CCATATTCTTTGCCTTTACCAG	630	[49]
sej	CATCAGAACTGTTGTTCCGCTAG	CTGAATTTTACCATCAAAGGTAC	142	[49]
sek	ACCGCTCAAGAGATTGAT	TTATATCGTTTCTTTATAAGAA	278	[50]
sel	AATATATAACTAGTGATCTAAAGGG	TATGGAATACTACACACCCCTTATA	359	[51]
sem	ATGCTGTAGATGTATATGGTCTAAG	CGTCCTTATAAGATATTTCTACATC	473	[51]
sen	ATGAGATTGTTCTACATAGCTGCAAT	AACTCTGCTCCCACTGAAC	680	[52]
seo	TGTAGTGTAAACAATGCATATGCAAATG	TTATGTAAATAAATAAACATCAATATGATGTC	722	[53]
sep	TTAGACAAACCTATTATCATAATGG	TATTATCATGTAACGTTACACCGCC	272	[53]
seq	AAGAGGTAACTGCTCAAG	TTATTCAGTCTTCTCATATG	285	[50]
ser	AAACCAGATCCAAGGCCTGGAG	TCACATTTGTAGTCAGGTGAACTT	700	[51]
ses	TTCAGAAATAGCCAATCATTTCAA	CCTTTTTGTTGAGAGCCGTC	195	[18]
set	GGTGATTATGTAGATGCTTGGG	TCGGGTGTTACTTCTGTTTGC	170	[18]
seu	TAAAATAAATGGCTCTAAAATTGATGG	ATCCGCTGAAAAATAGCATTGAT	141	[54]
sev	GCAGGATCCGATGTCGGAGTTTTG AATCTTAGG	TAACTGCAGTTAGTTACTATCTACATATGAT ATTTCGACATC	720	[55]
sew (*seu2*)	ATG TTA AAT GGC AAT CCT AAA C CA	TTA TTT TTT GGT TAA ATG AAC TTC TAC ATT AAT AGA TTT A		[21]

etiological agent. In addition, whatever the detection method used, it is crucial to concentrate the extract before performing detection assays due to the limited amount of SEs present in food. However, the food extract methods according to these commercial kits mainly include milk samples, ≥40% fat of food samples, ≥40% fat of food samples, and bacterial culture. The food matrix is very complex, and these extract methods are very difficult to meet the extraction and concentration of SEs. Therefore, it causes the low specificity of some commercial kits.

1.7 MASS SPECTROMETRY-BASED METHODS

Mass spectrometry (MS) possesses a great potential for qualitative and quantitative determination and analysis of SEs protein and peptide mixture.[66–69] MS is the currently most sensitive technique for the detection of SEs, as well as can rapidly provide a specific and reliable result. There are four soft ionization methods in MS including plasma desorption (PD), fast atom bombardment (FAB), electrospray ionization (ESI), and matrix-assisted laser desorption/ionization (MALDI). But SEs analysis mainly uses ESI[66–68] and MALDI.[69] A single MS technique cannot be used for all SEs proteins due to size limitation; however, MS has its advantage by combination with a series of other techniques. Currently, MS-based methods are widely used to detect SEs produced by *S. aureus*, including LC-ESI/MS,[66,67] online SPE-LC-ESI-MS/MS,[68] MALDI-TOF MS.

For food safety analysis, compositions of food samples can be very complicated. The food samples often contain many different proteins, lipids and many other molecular species that may interfere with the detection of the targeted SEs and then distort quantification. Thus, sample preparation remains a critical step for MS-based analysis. Till now, the procedures of sample preparation and treatment for MS-based analysis are only available for milk, fruit juices, and bacterial cultures. Taken together, these factors limit their usage in diagnostic analysis of SEs, and may affect their efficiency for analyzing other food samples.

1.8 CONCLUSIONS

SFP is one of the most common foodborne diseases, resulting from ingestion of SEs produced in food by enterotoxigenic strains of *S. aureus*.

The development of rapid, specific, sensitive, and reliable diagnostic approaches is not only important for detecting and distinguishing staphylococcal exterotoxins but also can provide powerful tool for food safety control in the public health and in food industry. Various methods, including animal tests, molecular-based methods, immunological methods, and MS technologies, have been applied in the determination and detection of SEs for food safety. Animal test cycle is too long and unable to meet the needs of real-time detection. PCR-based molecular biological analysis is a reliable and rapid approach with conclusive results. However, PCR-based methods can be used to detect SEs genes, but cannot determine the SEs at protein level. Immunological approaches such as ELISA and ELFA are able to directly and rapidly measure the levels of SEs in the food samples. Specific antibodies against each staphylococcal exterotoxin are required for the development of ELISA-based diagnostic kits. However, the combination of PCR and ELISA technologies enables us to detect the SE genes and measure the toxin levels in the food samples. The MS technologies are efficient and sensitive, but the requirements of tedious sample pretreatment steps, expensive instruments, and trained personnel limit their usage in the field. Every strategy has its own advantages and limitations for detection of SEs in the food samples. We have to choose different approaches if possible to determine the food contamination by staphylococcal exterotoxins.

KEYWORDS

- ***Staphylococcus aureus***
- **pathogen**
- **enterotoxin**
- **food poisoning**
- **contamination**
- **superantigens**

REFERENCES

1. Wang, X.; Meng, J.; Zhang, J.; Zhou, T.; Zhang, Y., et al. Characterization of *Staphylococcus aureus* Isolated from Powdered Infant Formula Milk and Infant Rice Cereal in China. *Int. J. Food Microbiol.* **2012,** *153,* 142–147.
2. Argudin, M. A.; Mendoza, M. C.; Rodicio, M. R. Food Poisoning and *Staphylococcus aureus* Enterotoxins. *Toxins (Basel).* **2010,** *2,* 1751–1773.
3. 1111 Review of Staphylococcal Food Poisoning in Hong Kong. Scientific Committee on Enteric Infections and Foodborne Diseases.
4. Wertheim, H. F. L.; Melles, D. C.; Vos, M. C.; van Leeuwen, W.; van Belkum, A., et al. The Role of Nasal Carriage in *Staphylococcus aureus* Infections. *Lancet Infect. Dis.* **2005,** *5,* 751–762.
5. Gorwitz, R. J.; Kruszon-Moran, D.; McAllister, S. K.; McQuillan, G.; McDougal, L. K., et al. Changes in the Prevalence of Nasal Colonization with *Staphylococcus aureus* in the United States, 2001–2004. *J. Infect. Dis.* **2008,** *197,* 1226–1234.
6. Fratamico, P. M.; Bhunia, A. K.; Smith, J. L. Foodborne Pathogens: Microbiology and Molecular Biology; Caister Academic Press: Norwich, England, 2005; p 334.
7. Hennekinne, J. A.; De Buyser, M. L.; Dragacci, S. *Staphylococcus aureus* and Its Food Poisoning Toxins: Characterization and Outbreak Investigation. *FEMS Microbiol. Rev.* **2012,** *36,* 815–836.
8. Hao, D.; Xing, X.; Li, G.; Wang, X.; Zhang, M., et al. Prevalence, Toxin Gene Profiles, and Antimicrobial Resistance of *Staphylococcus aureus* Isolated from Quick-Frozen Dumplings. *J. Food Prot.* **2015,** *78,* 218–223.
9. Cha, J. O.; Lee, J. K.; Jung, Y. H.; Yoo, J. I.; Park, Y. K., et al. Molecular Analysis of *Staphylococcus aureus* Isolates Associated with Staphylococcal Food Poisoning in South Korea. *J. Appl. Microbiol.* **2006,** *101,* 864–871.
10. Suzuki, Y.; Omoe, K.; Hu, D. L.; Sato'o, Y.; Ono, H. K., et al. Molecular Epidemiological Characterization of *Staphylococcus aureus* Isolates Originating from Food Poisoning Outbreaks that Occurred in Tokyo, Japan. *Microbiol. Immunol.* **2014,** *58,* 570–80.
11. Sato'o, Y.; Omoe, K.; Naito, I.; Ono, H. K.; Nakane, A., et al. Molecular Epidemiology and Identification of a *Staphylococcus aureus* Clone Causing Food Poisoning Outbreaks in Japan. *J. Clin. Microbiol.* **2014,** *52,* 2637–2640.
12. Wang, Y.; Wang, D.; Zhan, S.; Zheng, J.; Tao, Y., et al. Identification of *Staphylococcus aureus* and SpA Polymorphisms in Ma'anshan City. *J. Public Health Prevent. Med.* **2011,** *22,* 50–53.
13. Yan, X.; Wang, B.; Tao, X.; Hu, Q.; Cui, Z., et al. Characterization of *Staphylococcus aureus* Strains Associated with Food Poisoning in Shenzhen, China. *Appl. Environ. Microbiol.* **2012,** *78,* 6637–6642.
14. Hennekinne, J. A.; Ostyn, A.; Guillier, F.; Herbin, S.; Prufer, A. L., et al. How Should Staphylococcal Food Poisoning Outbreaks be Characterized? *Toxins (Basel).* **2010,** *2,* 2106–2116.
15. Balaban, N.; Rasooly, A. Staphylococcal Enterotoxins. *Int. J. Food Microbiol.* **2000,** *61,* 1–10.
16. Argaw, S.; Addis, M. A Review on Staphylococcal Food Poisoning. *Food Sci. Quality Manage.* **2015,** *40,* 59–72.

17. Argudin, M. A.; Mendoza, M. C.; Rodicio, M. R. Food Poisoning and *Staphylococcus aureus* Enterotoxins. *Toxins.* **2010,** *2,* 1751–1773.
18. Ono, H. K.; Omoe, K.; Imanishi, K.; Iwakabe, Y.; Hu, D. L., et al. Identification and Characterization of Two Novel Staphylococcal Enterotoxins, Types S and T. *Infect. Immun.* **2008,** *76,* 4999–5005.
19. Wilson, G. J.; Seo, K. S.; Cartwright, R. A.; Connelley, T.; Chuang-Smith, O. N., et al. A Novel Core Genome-encoded Superantigen Contributes to Lethality of Community-associated MRSA Necrotizing Pneumonia. *PLoS Pathog.* **2011,** *7,* e1002271.
20. Le Loir, Y.; Baron, F.; Gautier, M. *Staphylococcus aureus* and Food Poisoning. *Genet. Mol. Res.* **2003,** *2,* 63–76.
21. Thomas, D. Y.; Jarraud, S.; Lemercier, B.; Cozon, G.; Echasserieau, K., et al. Staphylococcal Enterotoxin-like Toxins U2 and V, Two New Staphylococcal Superantigens Arising from Recombination within the Enterotoxin Gene Cluster. *Infect. Immun.* **2006,** *74,* 4724–4734.
22. Bergdoll, M. S. Monkey Feeding Test for Staphylococcal Enterotoxin. *Methods Enzymol.* **1988,** *165,* 324–333.
23. Surgalla, M. J.; Bergdoll, M. S.; Dack, G. M. Some Observations on the Assay of Staphylococcal Enterotoxin by the Monkey-feeding Test. *J. Lab. Clin. Med.* **1953,** *41,* 782–788.
24. Scheuber, P. H.; Mossmann, H.; Beck, G.; Hammer, D. K. Direct Skin Test in Highly Sensitized Guinea Pigs for Rapid and Sensitive Determination of Staphylococcal Enterotoxin B. *Appl. Environ. Microbiol.* **1983,** *46,* 1351–1356.
25. Tollersrud, T.; Kampen, A. H.; Kenny, K. *Staphylococcus aureus* Enterotoxin D is Secreted in Milk and Stimulates Specific Antibody Responses in Cows in the Course of Experimental Intramammary Infection. *Infect. Immun.* **2006,** *74,* 3507–3512.
26. Barclay, R.; Ji, Y. Staphylococcal Toxins and Bovine Mastitis. *J. Vet. Sci. Med. Diagn.* **2014,** *4,*2.
27. Hu, D. L.; Omoe, K.; Shimoda, Y.; Nakane, A.; Shinagawa, K. Induction of Emetic Response to Staphylococcal Enterotoxins in the House Musk Shrew (*Suncus murinus*). *Infect. Immun.* **2003,** *71,* 567–570.
28. Ono, H. K.; Sato'o, Y.; Narita, K.; Naito, I.; Hirose, S., et al. Identification and Characterization of a Novel Staphylococcal Emetic Toxin. *Appl. Environ. Microbiol.* **2015,** *81,* 7034–7040.
29. Omoe, K.; Hu, D. L.; Ono, H. K.; Shimizu, S.; Takahashi-Omoe, H., et al. Emetic Potentials of Newly Identified Staphylococcal Enterotoxin-like Toxins. *Infect. Immun.* **2013,** *81,* 3627–3631.
30. Clark, W. G.; Page, J. S. Pyrogenic Responses to Staphylococcal Enterotoxins A and B in Cats. *J. Bacteriol.* **1968,** *96,* 1940–1946.
31. Tang, X.; Sun, R.; Hong, S.; Hu, G.; Yang, Y. Repeated Intranasal Instillation with Staphylococcal Enterotoxin B Induces Nasal Allergic Inflammation in Guinea Pigs. *Am. J. Rhinol. Allergy.* **2011,** *25,* 176–181.
32. Rasooly, R.; Do, P. M. In Vitro Cell-based Assay for Activity Analysis of Staphylococcal Enterotoxin A in Food. *FEMS Immunol. Med. Microbiol.* **2009,** *56,* 172–178.
33. Park, C. E.; Akhtar, M.; Rayman, M. K. Nonspecific Reactions of a Commercial Enzyme-linked Immunosorbent Assay Kit (TECRA) for Detection of Staphylococcal Enterotoxins in Foods. *Appl. Environ. Microbiol.* **1992,** *58,* 2509–2512.

34. Rasooly, R.; Do, P.; Hernlem, B. Sensitive, Rapid, Quantitative and In Vitro Method for the Detection of Biologically Active Staphylococcal Enterotoxin Type E. *Toxins (Basel).* **2016,** *8,* 150.
35. Rasooly, R.; Hernlem, B. TNF as Biomarker for Rapid Quantification of Active Staphylococcus Enterotoxin A in Food. *Sensors (Basel).* **2012,** *12.* 5978–5985.
36. Rasooly, L.; Rose, N. R.; Shah, D. B.; Rasooly, A. In Vitro Assay of *Staphylococcus aureus* Enterotoxin A Activity in Food. *Appl. Environ. Microbiol.* **1997,** *63,* 2361–2365.
37. Neill, R. J.; Fanning, G. R.; Delahoz, F.; Wolff, R.; Gemski, P. Oligonucleotide Probes for Detection and Differentiation of *Staphylococcus aureus* Strains Containing Genes for Enterotoxins A, B, and C and Toxic Shock Syndrome Toxin 1. *J. Clin. Microbiol.* **1990,** *28,* 1514–1518.
38. Liu, Y.; Zhang, J.; Ji, Y. PCR-based Approaches for the Detection of Clinical Methicillin-resistant *Staphylococcus aureus*. *Open Microbiol. J.* **2016,** *10,* 45–56.
39. Ikeda, T.; Tamate, N.; Yamaguchi, K.; Makino, S. Quantitative Analysis of *Staphylococcus aureus* in Skimmed Milk Powder by Real-time PCR. *J. Vet. Med. Sci.* **2005,** *67,* 1037–1041.
40. Klotz, M.; Opper, S.; Heeg, K.; Zimmermann, S. Detection of *Staphylococcus aureus* Enterotoxins A to D by Real-time Fluorescence PCR Assay. *J. Clin. Microbiol.* **2003,** *41,* 4683–4687.
41. Masoud, W.; Vogensen, F. K.; Lillevang, S.; Abu Al-Soud, W.; Sorensen, S. J., et al. The Fate of Indigenous Microbiota, Starter Cultures, *Escherichia coli*, *Listeria innocua* and *Staphylococcus aureus* in Danish Raw Milk and Cheeses Determined by Pyrosequencing and Quantitative Real Time (qRT)-PCR. *Int. J. Food Microbiol.* **2012,** *153,* 192–202.
42. Handayani, L.; Faridah, D. N.; Kusumaningrum, H. D. Staphylococcal Enterotoxin a Gene-carrying *Staphylococcus aureus* Isolated from Foods and its Control by Crude Alkaloid from Papaya Leaves. *J. Food Prot.* **2014,** *77,* 1992–1997.
43. Mu, X. Q.; Liu, B. B.; Hui, E.; Huang, W.; Yao, L. C., et al. A Rapid Loop-mediated Isothermal Amplification (LAMP) Method for Detection of the Macrolide-streptogramin Type B Resistance Gene *msrA* in *Staphylococcus aureus*. *J. Glob. Antimicrob. Resist.* **2016,** *7,* 53–58.
44. Suwanampai, T.; Pattaragulvanit, K.; Pattanamahakul, P.; Sutheinkul, O.; Okada, K., et al. Evaluation of Loop-mediated Isothermal Amplification Method for Detecting Enterotoxin a Gene of *Staphylococcus aureus* in Pork. *Southeast Asian J. Trop. Med. Public Health.* **2011,** *42,* 1489–1497.
45. Yin, H. Y.; Fang, T. J.; Wen, H. W. Combined Multiplex Loop-mediated Isothermal Amplification with Lateral Flow Assay to Detect Sea and Seb Genes of Enterotoxic *Staphylococcus aureus*. *Lett. Appl. Microbiol.* **2016,** *63,* 16–24.
46. Panneerseelan, L.; Muriana, P. M. An Immunomagnetic PCR Signal Amplification Assay for Sensitive Detection of *Staphylococcus aureus* Enterotoxins in Foods. *J. Food Prot.* **2009,** *72,* 2538–2546.
47. Chen, B.; Zheng, J.; Wang, Y.; Huang, X-r.; Lin, J., et al. Development of a Multiplex PCR-DHPLC Detection Method for Five Staphylococcal Enterotoxin Genes. *Food Sci.* **2014,** *35,* 243–248.

48. Martin, M. C.; Gonzalez-Hevia, M. A.; Mendoza, M. C. Usefulness of a Two-step PCR Procedure for Detection and Identification of Enterotoxigenic Staphylococci of Bacterial Isolates and Food Samples. *Food Microbiol.* **2003,** *20,* 605–610.
49. Peles, F.; Wagner, M.; Varga, L.; Hein, I.; Rieck, P., et al. Characterization of *Staphylococcus aureus* Strains Isolated from Bovine Milk in Hungary. *Int. J. Food Microbiol.* **2007,** *118,* 186–193.
50. Yarwood, J. M.; McCormick, J. K.; Paustian, M. L.; Orwin, P. M.; Kapur, V., et al. Characterization and Expression Analysis of *Staphylococcus aureus* Pathogenicity Island 3 - Implications for the Evolution of Staphylococcal Pathogenicity Islands. *J. Biol. Chem.* **2002,** *277,* 13138–13147.
51. Fueyo, J. M.; Mendoza, M. C.; Martin, M. C. Enterotoxins and Toxic Shock Syndrome Toxin in *Staphylococcus aureus* Recovered from Human Nasal Carriers and Manually Handled Foods: Epidemiological and Genetic Findings. *Microb. Infect.* **2005,** *7,* 187–194.
52. Jarraud, S.; Mougel, C.; Thioulouse, J.; Lina, G.; Meugnier, H., et al. Relationships between *Staphylococcus aureus* Genetic Background, Virulence Factors, Agr Groups (Alleles), and Human Disease. *Infect. Immun.* **2002,** *70,* 631–641.
53. Fueyo, J. M.; Mendoza, M. C.; Rodicio, M. R.; Muniz, J.; Alvarez, M. A., et al. Cytotoxin and Pyrogenic Toxin Superantigen Gene Profiles of *Staphylococcus aureus* Associated with Subclinical Mastitis in Dairy Cows and Relationships with Macrorestriction Genomic Profiles. *J. Clin. Microbiol.* **2005,** *43,* 1278–1284.
54. Letertre, C.; Perelle, S.; Dilasser, F.; Fach, P. Identification of a New Putative Enterotoxin SEU Encoded by the Egc Cluster of *Staphylococcus aureus*. *J. Appl. Microbiol.* **2003,** *95,* 38–43.
55. Fusco, V.; Quero, G. M.; Morea, M.; Blaiotta, G.; Visconti, A. Rapid and Reliable Identification of *Staphylococcus aureus* Harbouring the Enterotoxin Gene Cluster (egc) and Quantitative Detection in Raw Milk by Real Time PCR. *Int. J. Food Microbiol.* **2011,** *144,* 528–537.
56. Hait, J. M.; Tallent, S. M.; Bennett, R. W. Screening, Detection, and Serotyping Methods for Toxin Genes and Enterotoxins in Staphylococcus Strains. *J. AOAC Int.* **2014,** *97,* 1078–1083.
57. Peruski, A. H.; Johnson, L. H. 3rd; Peruski, L. F. Jr. Rapid and Sensitive Detection of Biological Warfare Agents Using Time-resolved Fluorescence Assays. *J. Immunol. Methods.* **2002,** *263,* 35–41.
58. Boynukara, B.; Gulhan, T.; Alisarli, M.; Gurturk, K.; Solmaz, H. Classical Enterotoxigenic Characteristics of *Staphylococcus aureus* Strains Isolated from Bovine Subclinical Mastitis in Van, Turkey. *Int. J. Food Microbiol.* **2008,** *125,* 209–211.
59. Kijek, T. M.; Rossi, C. A.; Moss, D.; Parker, R. W.; Henchal, E. A. Rapid and Sensitive Immunomagnetic-electrochemiluminescent Detection of Staphyloccocal Enterotoxin B. *J. Immunol. Methods.* **2000,** *236,* 9–17.
60. Aitichou, M.; Henkens, R.; Sultana, A. M.; Ulrich, R. G.; Sofi Ibrahim, M. Detection of *Staphylococcus aureus* Enterotoxin A and B Genes with PCR-EIA and a Handheld Electrochemical Sensor. *Mol. Cell. Probes.* **2004,** *18,* 373–377.
61. Su, Y. C.; Wong, A. C. Detection of Staphylococcal Enterotoxin H by an Enzyme-linked Immunosorbent Assay. *J. Food Prot.* **1996,** *59,* 327–330.

62. Omoe, K.; Ishikawa, M.; Shimoda, Y.; Hu, D. L.; Ueda, S., et al. Detection of Seg, Seh, and Sei Genes in *Staphylococcus aureus* Isolates and Determination of the Enterotoxin Productivities of *S. aureus* Isolates Harboring Seg, Seh, or Sei Genes. *J. Clin. Microbiol.* **2002,** *40,* 857–862.
63. Ikeda, T.; Tamate, N.; Yamaguchi, K.; Makino, S. Mass Outbreak of Food Poisoning Disease Caused by Small Amounts of Staphylococcal Enterotoxins A and H. *Appl. Environ. Microbiol.* **2005,** *71,* 2793–2795.
64. Zhao, Y.; Zhu, A.; Tang, J.; Tang, C.; Chen, J., et al. Identification and Measurement of Staphylococcal Enterotoxin-like Protein I (SEll) Secretion from *Staphylococcus aureus* Clinical Isolate. *J. Appl. Microbiol.* **2016,** *121,* 539–546.
65. Schlievert, P. M.; Case, L. C. Molecular Analysis of Staphylococcal Superantigens. *Methods Mol. Biol.* **2007,** *391,* 113–126.
66. Sospedra, I.; Soler, C.; Manes, J.; Soriano, J. M. Rapid Whole Protein Quantitation of Staphylococcal Enterotoxins A and B by Liquid Chromatography/Mass Spectrometry. *J. Chromatogr A* **2012,** *1238,* 54–59.
67. Dupuis, A.; Hennekinne, J. A.; Garin, J.; Brun, V. Protein Standard Absolute Quantification (PSAQ) for Improved Investigation of Staphylococcal Food Poisoning Outbreaks. *Proteomics.* **2008,** *8,* 4633–4636.
68. Andjelkovic, M.; Tsilia, V.; Rajkovic, A.; De Cremer, K.; Van Loco, J. Application of LC-MS/MS MRM to Determine Staphylococcal Enterotoxins (SEB and SEA) in Milk. *Toxins (Basel).* **2016,** *8,* 118.
69. Schlosser, G.; Kacer, P.; Kuzma, M.; Szilagyi, Z.; Sorrentino, A., et al. Coupling Immunomagnetic Separation on Magnetic Beads with Matrix-assisted Laser Desorption Ionization-time of Flight Mass Spectrometry for Detection of Staphylococcal Enterotoxin B. *Appl. Environ. Microbiol.* **2007,** *73,* 6945–6952.

PART II
Foodborne Bacterial Pathogens

CHAPTER 2

SALMONELLA SPECIES

LAN HU* and BAOGUANG LI

Division of Molecular Microbiology, OARSA, CFSAN, FDA, Laurel 20708, MD, USA

**Corresponding author. E-mail: lan16686@yahoo.com*

CONTENTS

Abstract..26
2.1 Introduction..26
2.2 Biological Characteristics..27
2.3 Epidemiology..28
2.4 Diseases..30
2.5 Genomics..33
2.6 Pathogenesis..37
2.7 Identification and Detection..38
2.8 Treatment and Prevention..41
2.9 Future Challenging..43
Acknowledgments..44
Keywords..44
References..44

ABSTRACT

Salmonella have been recognized as major foodborne pathogens and remain a tremendous medical and economic burden in the world. *Salmonella* enteric infections cause gastroenteritis, enteric fever, bacteremia, invasive nontyphoid *Salmonella* (NTS) disease, and asymptomatic carrier state. The most common serovars causing human and animal diseases are *Salmonella typhi*, *Salmonella paratyphi*, *Salmonella enteriditis*, *Salmonella typhimurium*, and *Salmonella choleraesuis*. It is important to understand the latest *Salmonella* epidemiological distribution, pathogenesis, genomics, and detection and identification methods. In this chapter, we will focus on the *Salmonella* epidemiology, genomics, molecular and biological detection methods, and the practical control measures.

2.1 INTRODUCTION

Salmonella species are Gram-negative, usually motile, facultative anaerobe, non-spore-forming, and rod-shaped bacteria. *Salmonella* have been recognized as major foodborne pathogens for humans and animals and remain a tremendous medical and economic burden. Most human pathogenic *Salmonellae* belong to *Salmonella enterica* subsp. *enterica* (*S. enterica*), which are divided into more than 2600 serovars (serotypes). The most common *Salmonella* serovars that cause human diseases include *Salmonella typhi*, *Salmonella enteritidis*, *Salmonella paratyphi*, *Salmonella typhimurium*, and *Salmonella choleraesuis*.[1,2]

Salmonella enteric infections result in diverse clinical manifestations including gastroenteritis, enteric fever (typhoid and paratyphoid fever), bacteremia, invasive non-typhoid *Salmonella* (iNTS) disease, and asymptomatic carrier state. The diseases caused by *Salmonellae* are termed Salmonellosis. Typhoid fever and paratyphoid fever are serious forms of Salmonellosis. The annual incidence of typhoid fever is estimated to be more than 27 million cases worldwide with 217,000 deaths, most of which occur in Asia and Africa.[3] Infections caused by nontyphoid *Salmonella* (NTS) (*S. enterica* serovars other than *S. typhi* or *S. paratyphi*) are estimated to cause approximately 93.8 million cases of enteric infections worldwide with 155,000 deaths each year.[4] In the United States, approximately 1.4 million cases with about 19,000 hospitalizations and about 500 deaths occur due to Salmonellosis annually.[5]

After vaccines against *S. typhi* and antibiotics are extensively used, the incidence of typhoid fever has been decreased. However, enteric fever continues to be a major infectious disease worldwide due to poor sanitation, the emerging of multiple drug-resistant strains, and lacking antibiotics/vaccines in the developing countries. Adding insult to injury, there is no vaccine available against paratyphoid fever. Gastroenteritis caused by NTS still is a major infectious disease worldwide. Thus, gaining an understanding of the latest *Salmonella* epidemiological distribution, pathogenesis, genomics, and detection and identification methods is a necessity for preventing and managing Salmonellosis. In this chapter, we will focus on the *Salmonella* epidemiology, genomics, molecular and biological detection methods, and the practical control measures.

2.2 BIOLOGICAL CHARACTERISTICS

The genus *Salmonella* consists of two species, *S. enterica* and *Salmonella bongori*. *S. enterica* is further divided into six subspecies including *S. enterica* subsp. *enterica*, *S. enterica* subsp. *salamae*, *S. enterica* subsp. *arizonae*, *S. enterica* subsp. *diarizonae*, *S. enterica* subsp. *houtenae*, and *S. enterica* subsp. *indica*.[1] Almost all *Salmonella* that cause disease in humans and domestic animals belong to *S. enterica* subsp. *enterica*.[1]

Salmonellae are also classified into more than 67 serogroups and 2600 serovars based on three surface antigens including lipopolysaccharide (LPS) "O" antigens, flagellar "H" antigens, and polysaccharide "Vi" capsular antigens.[1,2] The "O" antigens are found in all *Enterobacteriaceae* and are important for determining serotypes. The "H" antigens, which express two different phases (1 and 2), are specific. *S. typhi*, *S. paratyphi* C, and *Salmonella* Dublin express "Vi" capsule antigens, which is a superficial antigen overlying the "O" antigen and block O-agglutination in primary isolation. The most common serovars causing human and animal diseases are *S. typhi*, *S. paratyphi*, *Salmonella enteriditis*, *S. typhimurium*, and *S. choleraesuis*.[6] Seven polyvalent O antiserum mixtures including polyvalent O antiserum Group A, B, C, D, E, F, G, H, and L are available commercially to identify the serovars, and the antiserums can detect approximately 98% of the *Salmonella* isolates from humans and warm-blooded animals.[7]

Salmonellae are facultative anaerobe, intracellular bacteria. The size of *Salmonella* is 0.7–1.5 μm in the width and 2.0–5.0 μm in the length. The usual habitats for *S. enterica* are humans and warm-blooded animals. *Salmonellae* ferment glucose, but usually not lactose or sucrose, and all *Salmonellae* except for *S. typhi*, produce gas during the fermentation process. All strains in this genus are able to utilize citrate as a sole carbon source. The optimum growth temperature of *Salmonellae* is 35–37°C, but they can grow between 5 and 47°C. The optimal pH for *Salmonellae* growth is 6.5–7.5 with possibilities for growth at the pH values ranging from 4.5 to 9.0.[8,9] *Salmonellae* can survive in low-water activity (aw < 0.85) food or food ingredients for long time.[10] They are resistant to desiccation and may tolerate as much as a 20% salt for weeks. Under desiccation, *Salmonella* may change into viable but nonculturable (VBNC) state.[11–13] The VBNC may contribute to *Salmonella*'s long-term survive in harsh or stressful growth environments. The long-term survival capability plus the high virulence makes *Salmonella* a major challenge for food processing and storage.[14]

2.3 EPIDEMIOLOGY

2.3.1 INCIDENCE

S. typhi and *S. paratyphi* A, B, and C are limited in human only. Although enteric fever rarely occurs in Western countries, it is highly endemic especially in Asia and Africa. The mortality due to NTS infection is primarily restricted to the developing world. Invasive NTS causes a systemic, extraintestinal disease, which accounts for about 5% of NTS infections and occurs more commonly in Asia and sub-Saharan Africa. The incidence of iNTS disease is estimated approximately 3.4 million annually with about 681,316 deaths.[15] The most common *Salmonella* serovars caused iNTS diseases are *S. enteritidis* (33.1%) and *S. typhimu*rium (62.5%).[16,17] The epidemiology of iNTS disease in Africa is closely associated with malaria and malnutrition among infants and children, and about 95% of adult cases are immunosuppressed individuals such as HIV patients.[15,16,18] In the United States, NTS is a leading cause of gastroenteritis. Although *Salmonella* outbreaks are attributed to different *Salmonella* serovars, *S. typhimurium* and *S. enteritidis* are the most popular pathogens.[19,20]

2.3.2 INFECTIOUS ROUTES

Because *S. typhi* is restricted to humans, Typhoid fever is typically acquired by ingesting food or water contaminated by *S. typhi*-infected individuals or asymptomatic carriers. Food handlers who happen to be asymptomatic typhoid carriers can shed large number of *Salmonella* in their feces. "Typhoid Mary" is the classic example and the first known case of asymptomatic typhoid carriers who worked as a cook and infected 26–54 people in the New York City area during 1900–1915.[21]

NTS strains are found in the intestinal tract of many animals including poultry (e.g., chickens, ducks, geese, and turkeys), cows, pigs, and others. In developed countries, most NTS infected individuals are caused by ingestion of contaminated food or water.[10] *Salmonella* outbreaks are most commonly associated with uncooked poultry (30%), pork (8%), beef (8%), eggs (20%), and seafood.[22] Another important source of the organisms and recent outbreaks are associated with consuming contaminated vegetables and fruits, especially sprouts, tomatoes, and spinach.[23,24] The presence of *Salmonella* has been found in numerous dried foods or processed products, such as peanut butter, whole egg powder, cereals, herbs, soy bean meal, spices, and dried mushrooms.[25]

In some cases of *Salmonella* infections, NTS may be transmitted from person to person or from contact with pets such as cats, dogs, rodents, reptiles, or amphibians.[26,27] Although long-term carriage of NTS has not been reported, adults after *Salmonella* infection shed NTS organisms in their feces on average for a month and children under the age of 5 years after the infection shed the bacteria longer than one month.[28,29] Some strains of *Salmonella* can persistently survive in environments such as in soil and water, or on plants by producing biofilms.[30–32] Some *Salmonella* can even invade and hide inside certain edible plants.[33,34]

2.3.3 INFECTIOUS DOSE

The infectious dose varies among the *Salmonellae* serovars. For NTS, the infectious dose is approximately 10^3 bacteria by ingestion.[35] For enteric fever, the infectious dose is about 10^5 bacteria.[36] Individuals that suffer immunocompromised diseases or the elders may become infected with lower infectious doses. Over-the-counter medicines such as anti-acid drugs or food like milk may reduce the infectious doses by reducing the acidity of the stomach.

2.3.4 SUSCEPTIBLE POPULATIONS

Children under the age of five have higher rates of *Salmonella* infection than any other age group.[37] Elderly individuals and patients with chronic disease or who are immunocompromised have higher incidences than healthy individuals, and are also at high risk for serious outcome of Salmonellosis.[38,39]

2.3.5 ANTIBIOTIC RESISTANCE

Due to the extensive use of antibiotics in human medicine, animal feeds, and agricultural practice,[40] multiple antibiotic resistance *Salmonella* strains have emerged which are resistant to chloramphenicol, ampicillin, trimethoprim-sulfamethoxazole, spectrum cephalosporin, and others. The multiple antimicrobial resistance strains occur more often in NTS than in typhoid strains.

2.3.6 PEAK SEASONS

The peaks of Salmonellosis are summer and fall as is the cases of most enteric organisms.[41]

2.4 DISEASES

The clinical manifestations of *Salmonella* infections depend on host susceptibility and the type of *Salmonella* serovars involved.[5,37] The most common diseases caused by *Salmonellae* are gastroenteritis, enteric fever, bacteremia, invasive NTS disease, and asymptomatic carrier state.

S. typhi and *S. paratyphi* cause enteric fevers in human, and *S. typhimurium*, *S. dublin*, and *S. choleraesuis* cause disease in both humans and animals. However, these bacteria may cause a variety of symptoms in different hosts; for example, *S. typhimurium* causes gastroenteritis in humans but causes a typhoid-like systemic illness in mice. *S. enteritidis* is the most commonly identified pathogen in myocarditis[42] and *S. typhimurium*, *S. enteritidis*, *S. dublin*, and *S. choleraesuis* cause more invasive, extra-intestinal infections than any other strains.[19]

2.4.1 GASTROENTERITIS

The intestinal symptoms of enteric fever are not severe, sometimes are difficult to distinguish it from other diseases. In contrast to enteric fever, individuals infected with NTS have self-limiting, acute gastroenteritis and watery diarrhea. The common symptoms include nausea, vomit, abdominal pain, and fever. These symptoms usually appear 6–12 h after infection of *Salmonellae* and may last up to 10 days.[43]

2.4.2 ENTERIC FEVER

During infection, *S. typhi* and *S. paratyphi* penetrate the human small intestinal mucosa, internalize and survive within the mononuclear phagocytes, then disseminate via the bloodstream and cause extra-intestine infections. The incubation period for typhoid fever is 7–21 days and the symptoms persist for up to 3–5 weeks. In classic enteric fever, the onset of malaise, weakness, headache, and fever is slow and insidious. The fever rises gradually, is usually higher at night, until it reaches 39–40°C. Other frequent symptoms include chills, abdominal pain, dry cough, diarrhea, or constipation. During the second week of typhoid fever, about half of all cases exhibit "rose spots" or bacterial emboli on the chest, abdomen, and back.[44,45] Local tissue damage may occur due to the combined effects of ischemia (lack oxygen supply) and presence of bacteria. Intestinal perforation is a serious complication. Typhoid fever can be fatal if untreated and generally causes mortality in 5–30% of typhoid-infected individuals in the developing world. During the first week of acuter infection, bacteria can be cultured from stool in about 20% of cases and from blood in appropriate 75% of cases.

S. paratyphi A, B, and C also can cause enteric fever termed "paratyphoid fever," but the symptoms may be mild or may not distinguish from typhoid fever.[46,47] Although multi-resistant *S.typhi* has been the main cause of typhoid fever in Asia, the drug-resistant *S. paratyphi* A is emerging to cause more infections, and the incidence of paratyphoid fever is approximately 5.4 million cases per year.[48] The increased incidence of paratyphoid fever may have resulted from the successful use of several vaccines against *S. typhi* and/or the lack of vaccines against *S. paratyphi*.

2.4.3 BACTEREMIA AND iNTS DISEASES

Although the infection caused by NTS serovars usually are self-limiting, diarrhea with secondary bacteremia may occur in 3–10% of individuals, especially in the immunocompromised, children under 5 years of age, and patients with malaria, severe malnutrition, and anemia.[16] Bacteremia can result in septic shock, endocarditis, and secondary infection in organs such as bone marrow, liver, spleen, and gallbladder.

In developing countries, up to 5% NTS cases may be invasive, extra-intestinal disease. The symptoms of iNTS infections mimic those of enteric fever, and the symptoms usually observed are high fever, hepatosplenomegaly, respiratory complications, and often lacking intestinal symptoms. Invasive NTS disease has a high mortality rate of about 20–25% in Sub-Saharan Africa.[16]

2.4.4 THE ASYMPTOMATIC CARRIER STATE

Fecal excretion of *S. typhi* and *paratyphi* following acute illness is common, and may continue for weeks or months. An asymptomatic carriage state is used to describe those patients who shed *Salmonella* in feces or urine after the acute stage while the chronic carrier state is used to state the patients who excrete *Salmonella* in feces for more than one year. The carrier state occurs in 3–5% of the patients with enteric fever.[49] The asymptomatic carriage state is more common in children under the age of five, young adults, and elderly individuals. The chronic carrier state is often associated with gall bladder disease, which is more common in aged women and is public-health risk if they are food-handlers.[50] Carcinoma of the gall bladder is the fifth commonest gastrointestinal tract cancer in South America and Asia, and may be associated with the chronic typhoid carriage state.[51]

Asymptomatic carriage state of NTS occurs less frequently, however, it does happen in 0.5% adults and 3.9% children after the infection of NTS.[52] Antibiotics treatment does not prevent the shedding from these patients,[53] and may increase fecal excretion instead and prolong the carriage state because antimicrobial drugs may change the host immune response.[54]

2.5 GENOMICS

Multiple *Salmonella* strains have been sequenced and compared among *Salmonella* serovars and strains. The genome sizes of *S. typhi* Ty2, *S. typhi* CT18, and *S. typhimurium* LT2 are about 4.79, 4.86, and 4.81 megabases, respectively.[55–57] The genome size of *S. paratyphi* is smaller than most of *S. typhi* and *S. typhimurium* strains, for example, strain ATCC9150 and AKU-12601 are about 4.6 megabases.[58] These sequence data provide an accurate indication of genomic relationship among different *Salmonella* serovars and strains. About 90% of the genes between *S. typhi* and *S. typhimurium* serovars are identical;[57] however, the remained 10% genes make remarkable differences in pathogenic mechanism and host specialty between them.[59] Out of the 4400 genes that *S. typhi* and *S. paratyphi* share with *S. typhimurium*, about 200 genes in *S. typhi* and *S. paratyphi* are inactivated or degraded. Many of the degraded genes in the genomes of the enteric fever serovars are involved in motility, chemotaxis, and adhesion that associate with *S. typhimurium* pathogenesis.[58]

Salmonella virulence is associated with many factors encoded by genes that assist the organism to adhere, invade, survive, and multiply in the hosts. Understanding the genetic and molecular mechanisms will help to clarify *Salmonella* pathogenesis, host immunity, and the molecular basis of host specificity. Specific virulence factors of *Salmonella* include the *Salmonella* pathogenicity islands (SPIs), plasmids, flagella, fimbriae, outer membrane vesicles, and biofilm.

2.5.1 SPIs

Most *Salmonella* virulence factors are located on the SPIs. There are 21 SPIs known to date. SPI1 is approximately 40 kb in size and is located at 63 centisomes (cs) on the chromosome of *Salmonella*. SPI1 has an overall GC content of 42%, and contains approximate 30 genes encoded various components of a type 3 secretion system (T3SS)1 [119–123]. T3SS are molecular machines used to inject proteins (effectors) into eukaryotic host cells. The major proteins encoded by the genes of SPI1 and SPI2 are presented in Table 2.1. T3SS1 is involved in invading non-phagocytic cells and in activating proinflammatory responses.[60–62] SPI-2 locates at 30 cs on the chromosome and is a 25 kb in size. It encodes a T3SS2 and a regulatory system.[63–65] The T3SS2 is required for intramacrophage survival and systemic infection in mice. SPI3 (82 cs) is a 17 kb insertion at the *selC* tRNA locus, and encodes for *Salmonella*-specific *mgt*CB gene(s), which are required to survive in macrophages and to cause infections in mice.[66] SPI5 (20 cs) appears to mediate inflammation and chloride secretion.[67]

TABLE 2.1 The Major Function of the Common Proteins of T3SS in *S. enterica* Serovars.

Protein name	Acronym	Function
AvrA	Putative inner membrane protein	Inflammation, immune response
IagB	Invasion-associated gene	Invasion
InvC/InvE/InvJ	Invasion C/E/J	Invasion
OrgA/OrgB	Oxygen-regulated protein A/B	Bacterial internalization
PipB/PipB2	Pathogenicity island-encoded protein B/B2	Invasion
PrgH/PrgI/PrgJ	PhoP-repressed protein H/I/J	Virulence in mouse
SicA/SicP	*Salmonella* invasion chaperone A/P	Protein secretion, stabilization of SipB, SipC, and other proteins
SifA/SifB	*Salmonella*-induced filament A/B	Bacterial replication and bacterial survival
SipA, SipB, SipC, SipD	*Salmonella* invasion protein A/B/C/D	Invasion
SopA, SopB, SopE, SopE2	*Salmonella* outer protein A/B/E/E2	Invasion, chloride ion secretion
SlrP	*Salmonella* leucine-rich bacterial repeat protein	Bacterial survival
SpaK (*invB*), SpaL (*invC*), SpaM (*invI*), SpaN (invJ)	Surface presentation of antigen K/L/M/N	Invasion
SptP	*Salmonella* protein tyrosine	Bacterial survival
SseF, SseG, SseJ, SseL	*Salmonella* secreted effector F/G/J/L	Inhibition of immune response, bacterial survival
SspH1, SspH2	*Salmonella*-secreted protein	Inhibition of immune response, bacterial survival

Although the genomes of *S. typhimurium* and *S. typhi* share 11 common SPIs in 21 SPIs, SPI-7, SPI-15, SPI-17, and SPI-18 are present only in the genome of *S. typhi*, while SPI-14 is found only in *S. typhimurium*.[59] The differences may play a key in virulence and adherence factor composition, which affect host-pathogen interactions and disease outcome in the hosts.

2.5.2 OUTER MEMBRANE VESICLES (OMVs)

OMVs are spherical, bilayered structures that are naturally shed from the surfaces of Gram-negative bacteria, and contain LPS, phospholipids, and immunogenic proteins. These components can stimulate the innate immune system, and activate dendritic cells to present antigens.[68,69] In addition, the interaction of OMVs with *Salmonella*-specific T and B cells, involvement with pro-inflammatory response and antigenic presentation may make them potential vaccine candidates.[68]

2.5.3 POLYSACCHARIDES Vi CAPSULE

The polysaccharide Vi capsule is a key virulence factor of *S. typhi*.[70] The *via*A and *via*B genes are located on SPI-7 encoding for the biosynthesis of Vi capsule.[71] The Vi capsule can inhibit rapid neutrophil recruitment by blocking LPS recognition by pattern recognition receptors and by complement fragment C5a-dependent mechanism.[72,73] *S. typhi* does not cause a neutrophil influx in the intestine, but leads a disseminated systemic infection in human.

2.5.4 PhoPQ TWO-COMPONENT REGULATION SYSTEM

The PhoPQ two-component regulatory system plays a critical role in *Salmonella* pathogenesis. The system is required regulating virulence, surviving in macrophages and decreasing the response of the immune system. It also increases bacterial resistance to cationic antimicrobial peptides and assists the organisms surviving under low Mg^{2+} condition.[74]

2.5.5 ADHESION AND INVASION RELATED GENES

Genes involved in *Salmonella* adhesion and invasion in host cells include fimbriae (*pefA*), genes encoded effector proteins SopE, SopE2, SopB, and SipA of T3SS1,[75] hyper invasive locus (*hilA*).[76,77] *Salmonella* have 13 predicted fimbriae loci located on the bacterial surface, many of which are associated with attachment and colonization of host cells, and biofilm formation.[31,59,78,79]

Flagella are required for motility and chemotaxis. Motility contributes to *Salmonella* adherence and invasion of intestinal epithelial cells, and flagellin encoded by *fla* gene is an agonist of toll-like receptor (TLR)5, and potently activates the innate immune response.[80]

2.5.6 PLASMIDS

The plasmids of *Salmonella* strains carry some putative virulence-associated genes as well as those responsible for antimicrobial resistancwe. In *S. typhimurium*, a 90 kb virulent plasmid, named pSLT, has a highly conserved 8 kb *spv* locus, which consists of five ORF *spvRABCD*, is responsible for severe symptoms of Salmonellosis.[81,82] *spvB* is involved to intra-macrophage survival; but is absent in *S. typhi* and *S. paratyphi* A.[59,83] SopB is involved in invasion and AKT activation, which causes fluid secretion and *Salmonella* containing vacuole (SCV) formation.[79,84] Plasmids mediate antimicrobial resistant genes (against ampicillin, chloramphenicol, and trimethoprim-sulfamethoxazole) including *cat*, *dhfr7*, *dhfr14*, *sul1*, $bla_{\text{TEM-1}}$, and others.[59,85]

2.5.7 PHAGES

Several effector proteins T3SS such as SopE, SspH1, SseI, SodC-1, and SopE2 are encoded by phages or phage remnants. The phages can transfer virulence genes between different *Salmonella* strains.[86,87] Phages can be used for *Salmonella* typing. *S. typhimurium* phage type 104 (DT104) strain carries resistant genes against ampicillin, chloramphenicol, streptomycin, sulfonamides, and tetracycline.

2.5.8 BIOFILM-RELATED GENES

Salmonellae can form biofilm. *Salmonella* biofilm is composed of cellulose (encoded by *bcs* genes), thin fimbriae (encoded by *csg* genes), and BapA (encoded by *bap*A).[30,32,88–90,124] Biofilm formation is associated with the ability of *Salmonellae* to persist in hash environment and provide more resistance to antimicrobial agents than those growing in planktonic cultures. The genes related *Salmonella* biofilm formation and regulation are shown in Table 2.2.

TABLE 2.2 Some Genes Involved in Biofilm Formation and Regulation in *S. typhimurium*.

Gene	Encoded protein	Function
bcsA/B/C	Cellulose synthase catalytic subunit/cyclic di-GMP-binding protein/cellulose synthase operon protein	Cellulose synthesis
bcsE/F/G	Cellulose biosynthesis E/F/G	Cellulose assembly
csgA/B/C	Major curlin subunit/minor curlin subunit/curli assembly protein	Fimbrial curli synthesis and assembly
csgD	Transcriptional regulatory protein	Regulation of curli and other genes
csgE/F/G	Curli assembly protein E/F/G	Curli assembly
adrA	Diguanylate cyclase	Biofilm formation, colonization
bapA	Biofilm associated protein	Biofilm formation, colonization
sadA	A trimeric auto transporter adhesin	Biofilm formation, adhesion
ipfABCDE	Long polar fimbrial proteins	Adhesion, invasion, biofilm formation
pefA	Plasmid-encoded fimbriae	Adhesion, biofilm formation

2.6 PATHOGENESIS

After ingestion of contaminated food or water, *Salmonella* adhere and invade to the intestinal cells by using a T3SS1-mediated or with fimbrial/non-fimbrial adherence factors. After invasion or phagocytic uptake into epithelial cells or macrophages, *Salmonellae* remain within a modified phagosome known as SCV. Concomitantly, *Salmonella* T3SS2 is activated, facilitating pathogen survival and replication.[91] *Salmonellae* interact with conserved molecular patterns of the host cells such as TLRs and NOD-like receptors (NLRs). Bacterial flagellin interacts with TLR 5 [92] and the lipid A moiety of LPS interacts with TLR4/MD2/CD14 receptor complex,[93] while the O-antigen moiety of LPS interacts with complement component 3 fragment b (C3b).[94] The interaction of *Salmonella* "pathogen-associated-molecular-patterns" with the TLRs induces the production of proinflammatory cytokines and chemokines including interleukin (IL)-6, IL-1ß, tumor necrosis factor (TNF)-α, and these cytokines and chemokines recruit neutrophils and macrophages to the sites, and stimulate immune response.

Following ingestion, *S. typhi* proceeds to the ileum of the small intestine, passes the intestinal mucosa, adheres to and invades both epithelial cells and microfold (M) cells associated with Peyer's patches. This entry process involves specific host cell microfilament rearrangement and a signal transduction cascade. *S. typhi* prevents the attraction of neutrophils from inducing the local immune response. Subsequently, the pathogen survives and multiplies intracellularly within mononuclear phagocytes, and is spread systemically from the intestine to the mesenteric lymph, liver, spleen, bone marrow, and gallbladder, and causes a systemic infection.[9,19,95]

2.7 IDENTIFICATION AND DETECTION

Salmonella detection includes conventional culture methods, immunology-based assays, nucleic acid-based assays, matrix-associated laser desorption ionization-time of flight mass spectrometry (MALDI-TOF-MS), biosensors, and others. Conventional culture methods are laborious and time-consuming protocols, but still serve as the basis in clinical and food testing laboratories. Rapid, reliable, and sensible identification and detection of *Salmonella* in clinical samples, in food, and environmental screening are critical for public health and food safety.

2.7.1 TRADITIONAL CULTURE METHODS

Some samples (e.g., feces) can be plated directly onto highly selective and differential agars such as *Salmonella–Shigella* (SS) agar, xylose-lysine-deoxycholate (XLD) agar, xylose lactose tergitol 4 (XLT-4) agar, Hektoen enteric (HE), and MacConkey's agar plates. However, many types of specimens, such as food, blood, vomitus, and tissues may contain only a very low concentration of *salmonellae*, sometimes with large numbers of other bacteria, may need to pre-enrich and enrich in nonselective media, then culture in selective or differential media to form single colonies for identifying *Salmonella*.

2.7.2 TYPING AND SUBTYPING

A variety of *Salmonella* typing and subtyping methods has been developed. Phenotypic methods include serotyping and antimicrobial resistance typing, whereas genotypic methods include pulsed-field gel electrophoresis (PFGE), ribotyping, multilocus sequence typing (MLST), and others. In recent years, new molecular methods for *Salmonella* typing and subtyping have been developed using DNA-based assays to replace traditional serotyping. Sub-typing methods include PFGE, MLST, and other methods, which provide enhanced discriminatory power in the analysis of prevalent *Salmonella* serovars.

2.7.2.1 IMMUNOLOGY-BASED TYPING

The White–Kauffmann-antigenic scheme (WK) constitutes the principal diagnostic tool for the differentiation of more than 2600 *Salmonella* serovars. Currently, *Salmonella* isolates are classified using the WK scheme based on agglutination reactions with specific antisera detection of O (somatic), H (flagellar), and Vi antigens.[1] Because each isolate can be identified relatively easily and quickly by agglutination with antisera, WK serotyping is very useful in studying the epidemiology of *Salmonella* outbreaks. There are seven polyvalent O antiserum mixtures available commercially, and among them, OMA (polyvalent O antisera include group A, B, D, E, and L) and OMB (polyvalent O antisera include group C, F, G, and H) are used in the clinical setting to detect appropriately 98% of the *Salmonella* strains isolated from humans and warm-blooded animals. Despite its usefulness, serology-based serotyping is labor intensive and time consuming (3–5 days).[6,7] There are a number of commercially available, rapid serologic tests for Salmonellosis such as detecting anti-LPS IgM and IgG by ELISA, immune blot test, Widal test, and others.[96]

2.7.2.2 PULSED FIELD GEL ELECTROPHORESIS (PFGE)

Salmonella genomic DNA is digested by the restriction endonucleases such as XbaI, SmaI, etc., which generate a large number of DNA fragments. These DNA fragments are separated by agarose gel electrophoresis and analyzed with a software called Bionumerics.[97] The gels are stained with ethidium bromide and then visualized. The separated bands form what is now called a “DNA fingerprint pattern.” The fingerprint pattern from one strain can be compared with those of similar strains which are contained in a database called PulseNet at the Centers for Disease Control and Prevention. PFGE is the gold standard for investigation of *Salmonella* outbreaks and identification. After analysis, laboratories can upload patterns to the national database, where they can be investigated to see if the patterns form a new outbreak cluster or they are part of an ongoing outbreak.

2.7.2.3 MULTILOCUS SEQUENCE TYPING (MLST)

MLST is a typing method based on identifying internal nucleotide sequences of approximately 400–500 bp regions of multiple housekeeping genes. The unique sequence at each locus is assigned as an allele, and each unique combination of alleles (allelic profile) is specified a sequence type (ST).[98] Multiple related STs can then form clonal complexes. MLST analysis is used to infer evolutionary relationship among different *Salmonella* isolates and can be used to epidemiology and population genetics investigations.[99] The MLST data can be stored in database. At present, there are nine STs for *S. typhi* and seven STs for *S. partyphi*, and the STs of other serovars that have been documented in the MLST database (http://mlst.warwick.ac.uk/mlst/dbs/scentrica).

2.7.2.4 WHOLE GENOME SEQUENCING (WGS)

WGS provides the greatest resolution for microbial subtyping and is very useful for examination of bacterial evolution, source attribution, and transmission pathways during disease outbreaks. The WGS subtyping methods are becoming popular as the costs continue decreasing due to the development of next generation sequencing technology. The advantage of WGS is its reproducibility and high-throughput nature of the technology.[100,101]

2.7.3 PCR AND QUANTITATIVE PCR (qPCR)

Molecular methods include conventional polymerase chain reaction (PCR), real-time PCR, and others, which have been developed for the rapid and sensitive detection of *Salmonella*. The advantage of PCR is the potential to amplify small numbers of organisms, non-culture bacteria, and dead bacteria. If it is expected only living bacteria are detected, either choose the reverse-transcriptase-qPCR to detect RNA of living cells or use propidium monoazide in qPCR assay to enhance detection of live cells.[102,103] By and large, targets for *Salmonella* PCR assays are housekeeping genes and virulence genes such as the flagellin gene *fliC*, the capsular gene *viaB*, and other virulence-related *invA*, *sefC*, *hilA*, *sopB*, *stn*, *PefA*, *spvC* genes, and the 16s RNA gene.[103–105]

2.8 TREATMENT AND PREVENTION

2.8.1 TREATMENT

The majority of NTS infections are self-limiting. Because antibiotic therapy can increase the frequencies of producing drug-resistance and prolong the duration of excretion of bacteria, it is only recommended for people with severe illness and iNTS disease, or for high-risk groups including infants, the elder, and immunocompromised individuals.[19,29,106]

Enteric fever is always immediately treated with antibiotics. Due to rapid and widespread emergence of multi-drug resistant *S. typhi* and *S. paratyphi*, typhoid fever is becoming increasingly difficult to diagnose and treat. Multi-resistant strains also result in more severe disease outcome.[107–109]

The primary antimicrobial treatment options for Salmonellosis may include fluoroquinolones, ceftriaxone, azithromycin, and extended-spectrum cephalosporins.

2.8.2 VACCINES

The currently licensed *S. typhi* vaccines include a killed whole cell parenteral vaccine,[110] a live attenuated oral Ty21a,[111] and a typhoid Vi polysaccharide.[112] However, there are limitations with these vaccines in young children. For example, Ty21a could not be used in children under the age of 6 years in the United States (in children younger than 2 years old in other countries), and Vi polysaccharide vaccine could not be admitted under the age of 2 years. Although Ty21a provides some cross-protection against Paratyphoid B infection, no vaccines against *S. paratyphi* A are available.[113,114] The emergence of multidrug-resistant strains of *S. typhi* has added urgency and greater necessity to develop more effective typhoid vaccines.

Although vaccines against NTS serovars *S. enteritidis* and *S. typhimurium* are available for use in poultry,[115] no vaccines against NTS are available in humans or other animals. The development of vaccines against NTS and *S. paratyphi* A by using glyco-conjugate, live attenuated, and protein-based vaccine candidates is still underway.

2.8.3 AUTOPHAGY

Bacteriophages have specific host cell lysis enzymes called endolysins that can lyse bacterial hosts. The endolysins show no toxic to human cells.[116] Using bacteriophage and endolysins to biocontrol of *Salmonella* in food and food products show promise but are in their early phases of development.

2.8.4 ANTI-TUMOR THERAPIES

Salmonella-based anti-tumor immune-therapies deliver attenuated *S. typhimurium* into a tumor site to induce infiltration of neutrophils and tumor-specific T-cell responses, as well as reduce immunosuppressive cells. Inflammatory responses can be induced within the tumor microenvironment by delivery of attenuated *S. typhimurium*.[117,118]

2.8.5 PREVENTION

Salmonellosis can be prevented by breaking the transmission routes. However, to break the chain of spread of infection depends on adequate sewage disposal and clean water supplies, good food hygiene, good personal hygiene, rapid identification and efficient management of asymptomatic and chronic carriers, and provision of safe and effective vaccines to the high-risk populations.

Good food manufacturing practices and food handlers' hygiene are also a necessity to further reduce foodborne diseases. Such practices include adequate cooking, use of proper stocking temperatures and clean food processing equipment and work area, while good personal hygiene such as washing hands before eating and after using restroom is now highly advocated.

2.9 FUTURE CHALLENGES

Salmonellosis is still a tremendous burden on health care systems in the world. Therefore, controlling and preventing *Salmonella* infections remain a critical public health challenge.

Several questions regarding the bacterial-host interactions such as the progression of inflammation and immunity remain unanswered. For example, how can appropriate immune response against *Salmonellae* be triggered without causing excessive inflammation? Can we find common antigen(s) among different *Salmonella* serovars which could be used to develop vaccines?

Salmonella control measures such as adequate sewage disposal and clean water supplies have been very effective in reducing disease in the developed nations. At the same time, the effective inspection and surveillance of food and water should be strengthened to decrease the contamination of *Salmonella*.

Evidence of the development of herd immunity following the widespread use of the live, oral Ty21a typhoid vaccines suggests that an extensive immunization strategy in highly endemic areas might bring this disease under much better control. The emergence of multidrug-resistant strains of *Salmonellae* has added urgency to further develop effective and safe vaccines against NTS, *S. paratyphi* A, and *S. typhi*.

It is important to develop new approaches for advancing the detection and subtyping of *Salmonella* from food and environment sources, especially from fresh vegetables, fruits, meets, and ready-to-eat produces. The approaches include the molecular biological techniques that can rapidly, sensitively and specifically detect and identify the sources of contamination in food or water.

Over 100 years, antibiotics have saved uncountable human lives. However, due to the emergence of multi-drug resistant bacteria, new and effective antimicrobial drugs are needed to be developed. Yet, the safest ways to treat Salmonellosis may be the use of biocontrol ways such as bacteriophages, bacteriocins, herbs, and the substrates that can enhance innate immune response.

ACKNOWLEDGMENTS

The authors thank Dr. Ben D. Tall for critical review of the manuscript.

KEYWORDS

- ***Salmonella***
- **foodborne pathogen**
- **gastroenteritis**
- **typhoid fever**
- **invasive non-typhoid *Salmonella* (iNTS)**
- **biofilm**

REFERENCES

1. Grimont, P. A. D.; Weill, F. X. Antigenic Formulae of the *Salmonella* Serovars, 9th Revision World Health Organization Collaborating Center for Reference and Research on *Salmonella*; Pasteur Institute: Paris, France, 2007; pp 1–167.

2. Guibourdenche, M.; Roggentin, P.; Mikoleit, M.; Fields, P. I.; Bockemühl, J.; Grimont, P. A.; Weill, F. X. Supplement 2003–2007 (No. 47) to the White Kauffmann-Le Minor scheme. *Res. Microbiol.* **2010,** *161,* 26–29.
3. Crump, J. A.; Mintz, E. D. Global Trends in Typhoid and Paratyphoid Fever. *Clin. Infect. Dis.* **2010,** *50,* 241–246.
4. Majowicz, S. E.; Musto, J.; Scallan, E.; Angulo, F. J.; Kirk, M.; O'Brien, S. J.; Jones, T. F.; Fazil, A.; Hoekstra, R. M. International Collaboration on Enteric Disease 'Burden of Illness' Studies. The Global Burden of Non-typhoidal *Salmonella* Gastroenteritis. *Clin. Infect. Dis.* **2010,** *50,* 882–889.
5. Scallan, E.; Hoekstra, R. M.; Angulo, F. J.; Tauxe, R. V.; Widdowson, M. A.; Roy, S. L.; Jones, J. L.; Griffin, P. M. Foodborne Illness Acquired in the United States--major Pathogens. *Emerg. Infect. Dis.* **2011,** *17,* 7–15.
6. Su, L. H.; Chiu, C. H. *Salmonella*: Clinical Importance and Evolution of Nomenclature. *Chang Gung Med. J.* **2007,** *30,* 210–219.
7. Farmer, J. J.; Kelly, M. T. Enterobacteriaceae. In *Manual of Clinical Microbiology,* 5th ed.; Balows, A., Hausler W. J., Jr., Herrmann, K. L., Isenberg, H. D., Shadomy, H. J., Eds.; American Society for Microbiology: Washington, DC, 1991; pp 360–383.
8. Bryan, F. L.; Fanelli, M. J.; Riemann, H. *Salmonella* Infection. In *Food-Borne Infections and Intoxications,* 2nd ed.; Riemann, H., Bryan, F. L., Eds.; Academic Press: New York, NY, 1979; pp 73–130.
9. Hu, L.; Kopecko, D. J. Typhoid *Salmonella.* In *International Handbook of Foodborne Pathogens*; Miliotis M. D., Bier, J., Eds.; Marcel Dekker Inc.: New York, NY, 2003; Chapter 10, pp 151–67.
10. Beuchat, L. R.; Komitopoulou, E.; Beckers, H.; Betts, R. P.; Bourdichon, F.; Fanning, S.; Joosten, H. M.; Ter Kuile, B. H. Low-water Activity Foods: Increased Concern as Vehicles of Foodborne Pathogens. *J. Food Prot.* **2013,** *76,* 150–172.
11. Gruzdev, N.; Pinto, R.; Sela Saldinger, S. Persistence of *Salmonella enterica* during Dehydration and Subsequent Cold Storage. *Food Microbiol.* **2012,** *32,* 415–22.
12. Lesn, J.; Berthet, S.; Binard, S.; Rouxel, A.; Humbert, F. Changes in Culturability and Virulence of *Salmonella typhimurium* during Long-term Starvation under Desiccating Conditions. *Int. J. Food Microbiol.* **2000,** *60,* 195–203.
13. Podolak, R.; Enache, E.; Stone, W.; Black, D. G.; Elliott, P. H. Sources and Risk Factors for Contamination, Survival, Persistence, and Heat Resistance of *Salmonella* in Low-moisture Foods. *J. Food Prot.* **2010,** *73* (10), 1919–1936.
14. Roszak, D. B.; Grimes, D. J.; Colwell, R. R. Viable but Nonrecoverable Stage of *Salmonella enteritidis* in Aquatic Systems. *Can. J. Microbiol.* **1984,** *30,* 334–338.
15. Ao, T. T.; Feasey, N. A.; Gordon, M. A.; Keddy, K. H.; Angulo, F. J.; Crump, J. A. Global Burden of Invasive Nontyphoidal *Salmonella* Disease, 2010(1). *Emerg. Infect. Dis.* **2015,** *21* (6), 941–949.
16. Feasey, N. A.; Dougan, G.; Kingsley, R. A.; Heyderman, R. S.; Gordon, M. A. Invasive Non-typhoidal *Salmonella* Disease: An Emerging and Neglected Tropical Disease in Africa. *Lancet* **2012,** *379,* 2489–2499.
17. Reddy, E. A.; Shaw, A. V.; Crump, J. A. Community-acquired Bloodstream Infections in Africa: A Systematic Review and Meta-analysis. *Lancet Infect. Dis.* **2010,** *10,* 417–432.

18. Biggs, H. M.; Lester, R.; Nadjm, B.; Mtove, G.; Todd, J. E.; Kinabo, G. D.; Philemon, R.; Amos, B.; Morrissey, A. B.; Reyburn, H.; Crump, J. A. Invasive *Salmonella* Infections in Areas of High and Low Malaria Transmission Intensity in Tanzania. *Clin. Infect. Dis.* **2014,** *58,* 638–647.
19. Gordon, M. A.; Graham, S. M.; Walsh, A. L.; Wilson, L.; Phiri, A.; Molyneux, E.; Zijlstra, E. E.; Heyderman, R. S.; Hart, C. A.; Molyneux, M. E. Epidemics of Invasive *Salmonella enterica* Serovar *enteritidis* and *S. enterica* Serovar *typhimurium* Infection Associated with Multidrug Resistance among Adults and Children in Malawi. *Clin. Infect. Dis.* **2008,** *46,* 963–969.
20. Vojdani, J. D.; Beuchat, L. R.; Tauxe, R. V. Juice-associated Outbreaks of Human Illness in the United States, 1995 through 2005. *J. Food. Prot.* **2008,** *71,* 356–364.
21. Marr, J. S. Typhoid Mary. *Lancet* **1999,** *353,* 1714.
22. Gould, L. H.; Walsh, K.; Vieira, A. R.; Herman, K.; Williams, I. T.; Hall, A. J.; Cole. D. Surveillance for Foodborne Disease Outbreaks—United States, 1998–2008. *MMWR Surv. Summ.* **2013,** *62,* 1–34.
23. Bayer, C.; Bernard, H.; Prager, R.; Rabsch, W.; Hiller, P.; Malorny, B.; Pfefferkorn, B.; Frank, C.; de Jong, A.; Friesema, I.; Stark, K.; Rosner, B. An Outbreak of *Salmonella* Newport Associated with Mung Bean Sprouts in Germany and the Netherlands, October to November 2011. *Euro. Surveill.* **2014,** *19* (1), 20665.
24. Greene, S. K.; Daly, E. R.; Talbot, E. A.; Demma, L. J.; Holzbauer, S.; Patel, N. J.; Hill, T. A.; Walderhaug, M. O.; Hoekstra, R. M.; Lynch, M. F.; Painter, J. A. Recurrent Multistate Outbreak of *Salmonella* Newport Associated with Tomatoes from Contaminated Fields, 2005. *Epidemiol. Infect.* **2008,** *136* (2), 157–165.
25. Burgess, C. M.; Gianotti, A.; Gruzdev, N.; Holah, J.; Knøchel, S.; Lehner, A.; Margas, E.; Esser, S. S.; Sela Saldinger, S.; Tresse, O. The Response of Foodborne Pathogens to Osmotic and Desiccation Stresses in the Food Chain. *Int. J. Food Microbiol.* **2016,** *221,* 37–53.
26. Braden, C. R. *Salmonella enterica* Serotype *enteritidis* and Eggs: A National Epidemic in the United States. *Clin. Infect. Dis.* **2006,** *43,* 512–517.
27. Haeusler, G. M.; Curtis, N. Non-typhoidal *Salmonella* in Children: Microbiology, Epidemiology and Treatment. *Adv. Exp. Med. Biol.* **2013,** *764,* 13–26.
28. Buchwald D. S.; Blaser M. J. A Review of Human Salmonellosis: II. Duration of Excretion Following Infection with Nontyphi *Salmonella. Rev. Infect. Dis.* **1984,** *6,* 345–56.
29. Hohmann, E. L. Nontyphoidal Salmonellosis. *Clin. Infect. Dis.* **2001,** *32,* 263–269.
30. Austin, J. W.; Sanders, G.; Kay, W. W.; Collinson, S. K. Thin Aggregative Fimbriae Enhance *Salmonella* Enteritidis Biofilm Formation. *FEMS Microbiol. Lett.* **1998,** *162,* 295–301.
31. Ledeboer, N. A.; Frye, J. G.; McClelland, M.; Jones, B. D. *Salmonella enterica* Serovar *typhimurium* Requires the Lpf, Pef, and Tafi Fimbriae for Biofilm Formation on HEp-2 Tissue Culture Cells and Chicken Intestinal Epithelium. *Infect. Immun.* **2006,** *74,* 3156–3169.
32. Zogaj, X.; Nimtz, M.; Rohde, M.; Bokranz, W.; Romling, U. The Multicellular Morphotypes of *Salmonella typhimurium* and *Escherichia coli* Produce Cellulose as the Second Component of the Extracellular Matrix. *Mol. Microbiol.* **2001,** *39,* 1452–1463.

33. Arthurson, V.; Sessitsch, A.; Jäderlund, L. Persistence and Spread of *Salmonella enterica* Serovar Weltevreden in Soil and on Spinach Plants. *FEMS Microbiol. Lett.* **2011,** *314,* 67–74.
34. Golberg, D.; Kroupitski, Y.; Belausov, E.; Pinto, R.; Sela, S. *Salmonella typhimurium* Internalization is Variable in Leafy Vegetables and Fresh Herbs. *Int. J. Food Microbiol.* **2011,** *145,* 250–257.
35. Glynn, M. K.; Boop, C.; Dewitt, W.; Dabney, P.; Moktar, M.; Angulo, F. Emergence of Multidrug-resistant *Salmonella enterica* Serotype *typhimurium* DT104 Infections in the United States. *N. Engl. J. Med.* **1998,** *338,* 1333–1338.
36. Collins, C. H.; Kennedy, D. A. *Laboratory-acquired Infections*, 4th ed.; Butterworth-Heinermann: Oxford, UK, 1983.
37. *CDC. Foodborne Diseases Active Surveillance Network (FoodNet)*; FoodNet Surveillance Report for 2012 (Final Report), U.S. Department of Health and Human Services: Atlanta, Georgia, CDC, 2014.
38. *CDC. Suspecting Foodborne Illnesses in Special Populations*; Quick Facts for Providers, U.S. Department of Health and Human Services: Atlanta, Georgia, CDC, 2012.
39. Scallan, E.; Crim, S. M.; Runkle, A.; Henao, O. L.; Mahon, B. E.; Hoekstra, R. M.; Griffin, P. M. Bacterial Enteric Infections among Older Adults in the United States: Foodborne Diseases Active Surveillance Network, 1996–2012. *Foodborne Pathog. Dis.* **2015,** *12,* 492–499.
40. Graham, S. M. Nontyphoidal Salmonellosis in Africa. *Curr. Opin. Infect. Dis.* **2010,** *23* (5), 409–414.
41. Murray, P. R.; Baron, E. J.; Jorgensen, J. H.; Landry, M. L.; Pfaller, M. A. *Manual of Clinical Microbiology,* 9th ed.; ASM Press: Washington, DC, 2007.
42. Villablanca, P.; Mohananey, D.; Meier, G.; Yap, J. E.; Chouksey, S.; Abegunde, A. T. *Salmonella* Berta Myocarditis: Case Report and Systematic Review of Non-typhoid *Salmonella* Myocarditis. *World J. Cardiol.* **2015,** *7,* 931–937.
43. Glynn, J. R.; Palmer, S. R. Incubation Period, Severity of Disease, and Infecting Dose: Evidence from a *Salmonella* Outbreak. *Am. J. Epidemiol.* **1992,** *136,* 1369–1377.
44. Gilman, R. H.; Terminel, M.; Levine, M. M.; Hernandez-Mendoza, P.; Hornick, R. B. Relative Efficacy of Blood, Urine, Rectal Swab, Bone-marrow, and Rose-spot Cultures for Recovery of *Salmonella typhi* in Typhoid Fever. *Lancet* **1975,** *1,* 1211–1213.
45. Wangdi, T.; Winter, S. E.; Bäumler, A. J. Typhoid Fever: "You Can't Hit What You Can't See." *Gut Microbes.* **2012,** *3,* 88–92.
46. Meltzer, E.; Sadik, C.; Schwartz, E. Enteric Fever in Israeli Travelers: A Nationwide Study. *J. Travel Med.* **2005,** *12,* 275–281.
47. Patel, T. A.; Armstrong, M.; Morris-Jones, S. D.; Wright, S. G.; Doherty, T. Imported Enteric Fever: Case Series from the Hospital for Tropical Diseases, London, United Kingdom. *Am. J. Trop. Med. Hyg.* **2010,** *82,* 1121–1126.
48. Crump, J. A.; Luby, S. P.; Mintz, E. D. The Global Burden of Typhoid Fever. *Bull. World Health Organ.* **2004,** *82,* 346–353.
49. Caygill, C. P.; Braddick, M.; Hill, M. J.; Knowles, R. L.; Sharp, J. C. The Association between Typhoid Carriage, Typhoid Infection and Subsequent Cancer at a Number of Sites. *Eur. J. Cancer Prev.* **1995,** *4,* 187–193.

50. Gunn, J. S.; Marshall, J. M.; Baker, S.; Dongol, S.; Charles, R. C.; Ryan, E. T. *Salmonella* Chronic Carriage: Epidemiology, Diagnosis, and Gallbladder Persistence. *Trends Microbiol.* **2014,** *22,* 648–655.
51. Cruickshank, J. G.; Humphrey, T. J. The Carrier Food-handler and Non-typhoid Salmonellosis. *Epidemiol. Infect.* **1987,** *98,* 223–230.
52. Nagaraja, V.; Eslick, G. D. Systematic Review with Meta-analysis: The Relationship between Chronic *Salmonella typhi* Carrier Status and Gall-bladder Cancer. *Aliment Pharmacol. Ther.* **2014,** *39,* 745–750.
53. Murase, T.; Yamada, M.; Muto, T.; Matsushima, A.; Yamai, S. Fecal Excretion of *Salmonella enterica* Serovar *typhimurium* Following a Food-borne Outbreak. *J. Clin. Microbiol.* **2000,** *38* (9), 3495–3497.
54. Endt, K.; Stecher, B.; Chaffron, S.; Slack, E.; Tchitchek, N.; Benecke, A.; Van Maele, L.; Sirard, J. C.; Mueller, A. J.; Heikenwalder, M.; Macpherson, A. J.; Strugnell, R.; von Mering, C.; Hardt, W. D. The Microbiota Mediates Pathogen Clearance from the Gut Lumen after Non-typhoidal *Salmonella* Diarrhea. *PLoS Pathog.* **2010,** *6,* e1001097.
55. Deng, W.; Liou, S. R.; Plunkett, G. 3rd.; Mayhew, G. F.; Rose, D. J.; Burland, V.; Kodoyianni, V.; Schwartz, D. C.; Blattner, F. R. Comparative Genomics of *Salmonella enterica* Serovar Typhi Strains Ty2 and CT18. *J. Bacteriol.* **2003,** *185,* 2330–2237.
56. Parkhill, J.; Dougan, G.; James, K. D.; Thomson, N. R.; Pickard, D.; Wain, J.; Churcher, C.; Mungall, K. L.; Bentley, S. D.; Holden, M. T.; Sebaihia, M.; Baker, S.; Basham, D.; Brooks, K.; Chillingworth, T.; Connerton, P.; Cronin, A.; Davis, P.; Davies, R. M.; Dowd, L.; White, N.; Farrar, J.; Feltwell, T.; Hamlin, N.; Haque, A.; Hien, T. T.; Holroyd, S.; Jagels, K.; Krogh, A.; Larsen, T. S.; Leather, S.; Moule, S.; O'Gaora, P.; Parry, C.; Quail, M.; Rutherford, K.; Simmonds, M.; Skelton, J.; Stevens, K.; Whitehead, S.; Barrell, B. G. Complete Genome Sequence of a Multiple Drug Resistant *Salmonella enterica* Serovar *typhi* CT18. *Nature.* **2001,** *413,* 848–852.
57. McClelland, M.; Sanderson, K. E.; Clifton, S. W.; Latreille, P.; Porwollik, S.; Sabo, A.; Meyer, R.; Bieri, T.; Ozersky, P.; McLellan, M.; Harkins, C. R.; Wang, C.; Nguyen, C.; Berghoff, A.; Elliott, G.; Kohlberg, S.; Strong, C.; Du, F.; Carter, J.; Kremizki, C.; Layman, D.; Leonard, S.; Sun, H.; Fulton, L.; Nash, W.; Miner, T.; Minx, P.; Delehaunty, K.; Fronick, C.; Magrini, V.; Nhan, M.; Warren, W.; Florea, L.; Spieth, J.; Wilson, R. K. Complete Genome Sequence of *Salmonella enterica* Serovar *typhimurium* LT2. *Nature.* **2001,** *413,* 852–856.
58. McClelland, M.; Sanderson, K. E.; Clifton, S. W.; Latreille, P.; Porwollik, S.; Sabo, A.; Meyer, R.; Bieri, T.; Ozersky, P.; McLellan, M, Harkins CR, Wang C, Nguyen C, Berghoff A, Elliott G, Kohlberg S, Strong C, Du F, Carter J, Kremizki C, Layman D, Leonard S, Sun H, Fulton L, Nash W, Miner T, Minx P, Delehaunty K, Fronick C, Magrini V, Nhan M, Warren W, Florea L, Spieth J, Wilson RK. Comparison of Genome Degradation in *paratyphi* A and *typhi*, Human-restricted Serovars of *Salmonella enterica* that Cause Typhoid. *Nat. Genet.* **2004,** *36,* 1268–1274.
59. Sabbagh, S. C.; Forest, C. G.; Lepage, C.; Leclerc, J. M.; Daigle, F. So Similar, Yet so Different: Uncovering Distinctive Features in the Genomes of *Salmonella enterica* Serovars *typhimurium* and *typhi*. *FEMS Microbiol. Lett.* **2010,** *305,* 1–13.
60. Collazo, C. M.; Galán, J. E. The Invasion-associated Type-III Protein Secretion System in *Salmonella*–A Review. *Genes.* **1997,** *192,* 51–59.

61. Mills, D. M.; Bajaj, V.; Lee. C. A. A 40 Kilobase Chromosomal Fragment Encoding *Salmonella typhimurium* Invasion Genes is Absent from the Corresponding Region of the *Escherichia coli K-12* Chromosome. *Mol. Microbiol.* **1995,** *15,* 749–759.
62. Chen, L. M.; Kaniga, K.; Galán, J. E. *Salmonella* Spp. are Cytotoxic for Cultured Macrophages. *Mol. Microbiol.* **1996,** *21,* 1101–1115.
63. Ochman, H.; Soncini, F. C.; Solomon, F.; Groisman, E. A. Identification of a Pathogenicity Island Required for *Salmonella* Survival in Host Cells. *Proc. Natl. Acad. Sci. USA* **1996,** *93,* 7800–7804.
64. Shea, J. E.; Hensel, M.; Gleeson, C.; Holden. D. W. Identification of a Virulence Locus Encoding a Second Type III Secretion System in *Salmonella typhimurium. Proc. Natl. Acad. Sci. USA* **1996,** *93,* 2593–2597.
65. Hensel, M.; Shea, J. E.; Bäumler, A. J.; Gleeson, C.; Blattner, F.; Holden, D. W. Analysis of the Boundaries of *Salmonella* Pathogenicity Island 2 and the Corresponding Chromosomal Region of *Escherichia coli* K-12. *J. Bacteriol.* **1997,** *179,* 1105–1111.
66. Blanc-Potard, A. B.; Groisman, E. A. The *Salmonella selC* Locus Contains a Pathogenicity Island Mediating Intramacrophage Survival. *EMBO J.* **1997,** *16* (17), 5376–5385.
67. Wood, M. W.; Jones, M. A.; Watson, P. R.; Hedges, S.; Wallis, T. S.; Galyo, E. E. Identification of a Pathogenicity Island Required for *Salmonella* Enteropathogenicity. *Mol. Microbiol.* **1998,** *29,* 883–891.
68. Alaniz, R. C.; Deatherage, B. L.; Lara, J. C.; Cookson, B. T. Membrane Vesicles are Immunogenic Facsimiles of *Salmonella typhimurium* that Potently Activate Dendritic Cells, Prime B and T cell Responses, and Stimulate Protective Immunity In Vivo. *J. Immunol.* **2007,** *179,* 7692–7701.
69. Daleke-Schermerhorn, M. H.; Felix, T.; Soprova, Z.; Ten Hagen-Jongman, C. M.; Vikström, D.; Majlessi, L.; Beskers, J.; Follmann, F.; de Punder, K.; van der Wel, N. N.; Baumgarten, T.; Pham, T. V.; Piersma, S. R.; Jiménez, C. R.; van Ulsen, P.; de Gier, J. W.; Leclerc, C.; Jong, W. S.; Luirink, J. Decoration of Outer Membrane Vesicles with Multiple Antigens by Using an Autotransporter Approach. *Appl. Environ. Microbiol.* **2014,** *80,* 5854–5865.
70. Wain, J.; House, D.; Zafar, A.; Baker, S.; Nair, S.; Kidgell, C.; Bhutta, Z.; Dougan, G.; Hasan, R. Vi Antigen Expression in *Salmonella enterica* Serovar Typhi Clinical Isolates from Pakistan. *J. Clin. Microbiol.* **2005,** *43,* 1158–1165.
71. Kolyva, S.; Waxin, H.; Popoff, M. Y. The Vi Antigen of *Salmonella typhi*: Molecular Analysis of the ViaB Locus. *J. Gen. Microbiol.* **1992,** *138,* 297–304.
72. Wilson, R. P.; Raffatellu, M.; Chessa, D.; Winter, S. E.; Tükel, C.; Bäumler, A. J. The Vi-capsule Prevents Toll-like Receptor 4 Recognition of *Salmonella. Cell. Microbiol.* **2008,** *10,* 876–890.
73. Wangdi, T.; Lee, C. Y.; Spees, A. M.; Yu, C.; Kingsbury, D. D.; Winter, S. E.; Hastey, C. J.; Wilson, R. P.; Heinrich, V.; Bäumler, A. J. The Vi Capsular Polysaccharide Enables *Salmonella enterica* Serovar *typhi* to Evade Microbe-guided Neutrophil Chemotaxis. *PLoS Pathog.* **2014,** *10,* e1004306.
74. Dalebroux, Z. D.; Miller, S. I. *Salmonellae* PhoPQ Regulation of the Outer Membrane to Resist Innate Immunity. *Curr. Opin. Microbiol.* **2014,** *17,* 106–113.
75. Van Asten, A. J.; van Dijk, J. E. Distribution of Classic Virulence Factors among *Salmonella* Spp. *FEMS Immunol. Med. Microbiol.* **2005,** *44,* 251–259.

76. Lostroh, C. P.; Bajaj, V.; Lee, C. A. The Cis Requirements for Transcriptional Activation by HilA, a Virulence Determinant Encoded on SPI-1. *Mol. Microbiol.* **2000,** *37,* 300–315.
77. Murugkar, H. V.; Rahman, H.; Dutta, P. K. Distribution of Virulence Genes in *Salmonella* Serovars Isolated from Man and Animals. *Indian J. Med. Res.* **2003,** *117,* 66–70.
78. van der Velden, A. W.; Baumler, A. J.; Tsolis, R. M.; Heffron, F. Multiple Fimbrial Adhesins are Required for Full Virulence of *Salmonella typhimurium* in Mice. *Infect. Immun.* **1998,** *66,* 2803–2808.
79. Ibarra, J. A.; Steele-Mortimer, O. *Salmonella*--the Ultimate Insider. *Salmonella* Virulence Factors that Modulate Intracellular Survival. *Cell. Microbiol.* **2009,** *11* (11), 1579–1586.
80. Stecher, B.; Hapfelmeier, S.; Muller, C.; Kremer, M.; Stallmach, T.; Hardt, W. D. Flagella and Chemotaxis are Required for EfficientIinduction of *Salmonella enterica* Serovar Typhimurium Colitis in Streptomycin-pretreated Mice. *Infect. Immun.* **2004,** *72,* 4138–4150.
81. Guiney, D. G.; Fierer, J. The Role of the *spv* Genes in *Salmonella* Pathogenesis. *Front Microbiol.* **2011,** *2,* 129.
82. Browne, S. H.; Lesnick, M. L.; Guiney, D. G. Genetic Requirements for *Salmonella*-Induced Cytopathology in Human Monocyte-derived Macrophages. *Infect. Immun.* **2002,** *70,* 7126–7135.
83. Lesnick, M. L.; Reiner, N. E.; Fierer, J.; Guiney, D. G. The *Salmonella spvB* Virulence Gene Encodes an Enzyme that ADP-ribosylates Actin and Destabilizes the Cytoskeleton of Eukaryotic Cells. *Mol. Microbiol.* **2001,** *39,* 1464–1470.
84. Terebiznik, M. R.; Vieira, O. V.; Marcus, S. L.; Slade, A.; Yip, C. M.; Trimble, W. S.; Meyer, T.; Finlay, B. B.; Grinstein, S. Elimination of Host Cell PtdIns(4,5)P(2) by Bacterial SigD Promotes Membrane Fission during Invasion by *Salmonella. Nat. Cell Biol.* **2002,** *4,* 766–773.
85. Phan, M. D.; Kidgell, C.; Nair, S.; Holt, K. E.; Turner, A. K.; Hinds, J.; Butcher, P.; Cooke, F. J.; Thomson, N. R.; Titball, R.; Bhutta, Z. A.; Hasan, R.; Dougan, G.; Wain, J. Variation in *Salmonella enterica* Serovar *typhi* IncHI1 Plasmids during the Global Spread of Resistant Typhoid Fever. *Antimicrob. Agents Chemother.* **2009,** *53,* 716–727.
86. Ehrbar, K.; Hardt, W. D. Bacteriophage-encoded Type III Effectors in *Salmonella enterica* Subspecies 1 Serovar *typhimurium*. *Infect. Genet. Evol.* **2005,** *5,* 1–9.
87. Mirold, S.; Rabsch, W.; Tschäpe, H.; Hardt, W. D. Transfer of the *Salmonella* Type III Effector *sopE* between Unrelated Phage Families. *J. Mol. Biol.* **2001,** *312,* 7–16.
88. Solano, C.; García, B.; Valle, J.; Berasain, C.; Ghigo J. M.; Gamazo, C.; Lasa, I. Genetic Analysis of *Salmonella enteritidis* Biofilm Formation: Critical Role of Cellulose. *Mol. Microbiol.* **2002,** *43,* 793–808.
89. Biswas, R.; Agarwal, R. K.; Bhilegaonkar, K. N.; Kumar, A.; Nambiar, P.; Rawat, S.; Singh, M. Cloning and Sequencing of Biofilm-associated Protein (BapA) Gene and Its Occurrence in Different Serotypes of *Salmonella*. *Lett. Appl. Microbiol.* **2011,** *52,* 138–143.
90. Römling, U. Characterization of the Rdar Morphotype, a Multicellular Behaviour in *Enterobacteriaceae*. *Cell. Mol. Life Sci.* **2005,** *62,* 1234–1246.

91. Figueira, R.; Holden, D. W. Functions of the *Salmonella* Pathogenicity Island 2 (SPI-2) Type III Secretion System Effectors. *Microbiology* **2012,** *158,* 1147–1161.
92. Gewirtz, A. T.; Navas, T. A.; Lyons, S.; Godowski, P. J.; Madara, J. L. Cutting Edge: Bacterial Flagellin Activates Basolaterally Expressed TLR5 to Induce Epithelial Proinflammatory Gene Expression. *J. Immunol.* **2001,** *167,* 1882–1885.
93. Joiner, K. A.; Ganz, T.; Albert, J.; Rotrosen, D. The Opsonizing Ligand on *Salmonella typhimurium* Influences Incorporation of Specific, but not Azurophil, Granule Constituents into Neutrophil Phagosomes. *J. Cell Biol.* **1989,** *109,* 2771–2782.
94. Vazquez-Torres, A.; Vallance, B. A.; Bergman, M. A.; Finlay, B. B.; Cookson, B. T.; Jones-Carson, J.; Fang, F. C. Toll-like Receptor 4 Dependence of Innate and Adaptive Immunity to *Salmonella*: Importance of the Kupffer Cell Network. *J. Immunol.* **2004,** *172,* 6202–6208.
95. Pascopella, L.; Raupach, B.; Ghori, N.; Monack, D.; Falkow, S.; Small, P. L. Host Restriction Phenotypes of *Salmonella typhi* and *Salmonella gallinarum. Infect. Immun.* **1995,** *63,* 4329–4335.
96. Crump, J. A.; Sjolund-Karlsson, M.; Cordon, M. A.; Parry, C. M. Epidemiology, Clinical Presentation, Laboratory Diagnosis, Antimicrobial Resistance, and Antimicrobial Management of Invasive *Salmonella* Infections. *Clin. Microbiol. Rev.* **2015,** *28,* 901–937.
97. Ribot, E. M.; Fair, M. A.; Gautom, R.; Cameron, D. N.; Hunter, S. B.; Swaminathan, B.; Barrett, T. J. Standardization of Pulsed-field Gel Electrophoresis Protocols for the Subtyping of *Escherichia coli* O157:H7, *Salmonella*, and *Shigella* for PulseNet. *Foodborne Pathog. Dis.* **2006,** *3,* 59–67.
98. Enright, M. C.; Spratt, B. G. Multilocus Sequence Typing. *Trends Microbiol.* **1999,** *7,* 482–487.
99. Achtman, M.; Wain, J.; Weill, F. X.; Nair, S.; Zhou, Z.; Sangal, V.; Krauland, M. G.; Hale, J. L.; Harbottle, H.; Uesbeck, A.; Dougan, G.; Harrison, L. H.; Brisse, S.; Enterica, S. MLST Study Group. Multilocus Sequence Typing as a Replacement for Serotyping in *Salmonella enterica. PLoS Pathog.* **2012,** *8,* e1002776.
100. Kwong, J.; Mccallum, N.; Sintchenko, V.; Howden, B. Whole Genome Sequencing in Clinical and Public Health Microbiology. *Pathology* **2015,** *47,* 199–210.
101. Sabat, A. J.; Budimir, A.; Nashev, D.; Sá-Leão, R.; van Dijl, J. M.; Laurent, F.; Grundmann, H.; Friedrich, A. W. Overview of Molecular Typing Methods for Outbreak Detection and Epidemiological Surveillance. *Euro. Surveill.* **2013,** *18,* 20380.
102. Gonzalez-Escalona, N.; Hammack, T. S.; Russell, M.; Jacobson, A. P.; De Jesus, A. J.; Brown, E. W.; Lampel, K. A. Detection of Live *Salmonella* Sp. Cells in Produce by a Taqman-based Quantitative Reverse Transcriptase Real-Time PCR Targeting *invA* mRNA. *Appl. Environ. Microbiol.* **2009,** *75,* 3714–3720.
103. Li, B.; Chen, J. Q. Development of a Sensitive and Specific qPCR Assay in Conjunction with Propidium Monoazide for Enhanced Detection of Live *Salmonella* spp. in Food. *BMC Microbiol.* **2013,** *13,* 273.
104. Lee, K.; Iwata, T.; Shimizu, M.; Taniguchi, T.; Nakadai, A.; Hirota, Y.; Hayashidani, H. A Novel Multiplex PCR Assay for *Salmonella* Subspecies Identification. *J. Appl. Microbiol.* **2009,** *107,* 805–811.
105. Skyberg, J. A.; Logue, C. M.; Nolan, L. K. Virulence Genotyping of *Salmonella* spp. with Multiplex PCR. *Avian Dis.* **2006,** *50,* 77–81.

106. Forsythe, C. T.; Ernst, M. E. Do Fluoroquinolones Commonly Cause Arthropathy in Children? *CJEM.* **2007,** *9,* 459–462.
107. Centers for Disease Control and Prevention, US Department of Health and Human Services. Antibiotic Resistance Threats in the United States. Atlanta: CDC; 2013. Available from: http://www.cdc.gov/drugresistance/pdf/ar-threats-2013-508.pdf
108. Ochiai, R. L.; Wang, X. Y.; von Seidlein, L.; Yang, J.; Bhutta, Z. A.; Bhattacharya, S. K.; Agtini, M.; Deen, J. L.; Wain, J.; Kim, D. R, Ali M, Acosta CJ, Jodar L, Clemens JD. *Salmonella paratyphi* A Rates, Asia. *Emerg. Infect. Dis.* **2005,** *11,* 1764–1766.
109. Teh, C. S.; Chua, K. H.; Thong, K. L. Paratyphoid Fever: Splicing the Global Analyses. *Int. J. Med. Sci.* **2014,** *11* (7), 732–741.
110. Engels, E. A.; Falagas, M. E.; Lau, J.; Bennish, M. L. Typhoid Fever Vaccines: A Meta-analysis of Studies on Efficacy and Toxicity. *BMJ.* **1998,** *316,* 110–116.
111. Germanier, R.; Füer, E. Isolation and Characterization of Gal E mutant Ty 21a of *Salmonella typhi*: A Candidate Strain for a Live, Oral Typhoid Vaccine. *J. Infect. Dis.* **1975,** *131,* 553–558.
112. Tacket, C. O.; Ferreccio, C.; Robbins, J. B.; Tsai, C. M.; Schulz, D.; Cadoz, M.; Goudeau, A.; Levine, M. M. Safety and Immunogenicity of Two *Salmonella typhi* Vi Capsular Polysaccharide Vaccines. *J. Infect. Dis.* **1986,** *154,* 342–345.
113. Levine, M. M.; Ferreccio, C.; Black, R. E.; Lagos, R.; San Martin, O.; Blackwelder, W. C. Ty21a Live Oral Typhoid Vaccine and Prevention of Paratyphoid Fever Caused by *Salmonella enterica* Serovar *paratyphi* B. *Clin. Infect. Dis.* **2007,** *45* (Suppl. 1), S24–28.
114. Waddington, C. S.; Darton, T. C.; Pollard, A. J. The Challenge of Enteric Fever. *J. Infect.* **2014,** *68* (Suppl. 1), S38–50.
115. Desin, T. S.; Köster, W.; Potter, A. A. *Salmonella* Vaccines in Poultry: Past, Present and Future. *Expert. Rev. Vaccines.* **2013,** *12,* 87–96.
116. Borysowski, J.; Weber-Dabrowska, B.; Górski, A. Bacteriophage Endolysins as a Novel Class of Antibacterial Agents. *Exp. Biol. Med. (Maywood).* **2006,** *231,* 366–377.
117. MacLennan, C. A.; Martin, L. B.; Micoli, F. Vaccines against Invasive *Salmonella* Disease: Current Status and Future Directions. *Hum. Vaccin. Immunother.* **2014,** *10,* 1478–1493.
118. Hong, E. H.; Chang, S. Y.; Lee, B. R.; Pyun, A. R.; Kim, J. W.; Kweon, M. N.; Ko, H. J. Intratumoral Injection of Attenuated *Salmonella* Vaccine Can Induce Tumor Microenvironmental Shift from Immune Suppressive to Immunogenic. *Vaccine.* **2013,** *31,* 1377–1384.
119. Agbor, T. A.; McCormick, B. A. *Salmonella* Effectors: Important Players Modulating Host Cell Function During Infection. *Cell. Microbiol.* **2011,** *13,* 1858–1869.
120. Fàbrega, A.; Vila, J. *Salmonella enterica* Serovar *typhimurium* Skills to Succeed in the Host: Virulence and Regulation. *Clin. Microbiol. Rev.* **2013,** *26,* 308–341.
121. Galán, J. E. Common Themes in the Design and Function of Bacterial Effectors. *Cell Host Microbe.* **2009,** *5,* 571–579.
122. Ramos-Morales, F. Impact of *Salmonella* Enterica Type III Secretion System Effectors on the Eukaryotic Host Cell. *ISRN Cell Biol.* **2012,** *2012,* ID 787934.
123. The UniProt Consortium. UniProt: The Universal Protein Knowledgebase. *Nucleic Acids Res.* **2017,** *45,* D158–D169.

124. Raghunathan, D.; Wells, T. J.; Morris, F. C.; Shaw, R. K.; Bobat, S.; Peters, S. E.; Paterson, G. K.; Jensen, K. T.; Leyton, D. L.; Blair, J. M.; Browning, D. F.; Pravin, J.; Flores-Langarica, A.; Hitchcock, J. R.; Moraes, C. T.; Piazza, R. M.; Maskell D. J.; Webber, M. A.; May, R. C.; MacLennan, C. A.; Piddock, L. J.; Cunningham, A. F.; Henderson, I. R. SadA, a Trimeric Autotransporter from *Salmonella enterica* Serovar Typhimurium, can Promote Biofilm Formation and Provides Limited Protection against Infection. *Infect. Immun.* **2011,** *79,* 4342–4352.

CHAPTER 3

CAMPYLOBACTER SPECIES

LAN HU[1*] and DENNIS D. KOPECKO[2]

[1]*Division of Molecular Microbiology, OARSA, CFSAN, FDA, Laurel 20708, MD, USA*

[2]*Combivax, LLC, Silver Spring 20906, MD, USA*

[*]*Corresponding author. E-mail: lan16686@yahoo.com*

CONTENTS

Abstract 56
3.1 Introduction 56
3.2 Biological Characteristics 57
3.3 Epidemiology 59
3.4 Diseases 61
3.5 Genomics 64
3.6 Pathogenesis 70
3.7 Identification and Detection 73
3.8 Treatment and Prevention 78
3.9 Advances in *Campylobacter* Research 79
Keywords 81
References 82

ABSTRACT

Campylobacter spp. are the leading cause of foodborne gastroenteritis in humans worldwide. *Campylobacter* can also trigger serious post-infectious sequelae such as reactive arthritis and Guillain–Barré syndrome. It is important to understand the latest *Campylobacter* epidemiology, pathogenesis, genomics, identification, and detection methods. This chapter will present an overview of the general concepts, mechanisms, new identification and detection methods. We will focus on *Campylobacter jejuni* and *Campylobacter coli*, which are the main causes of *Campylobacter* enteritis in humans, to improve food safety as well as public health.

3.1 INTRODUCTION

Campylobacter (*C.*) species are small, microaerophilic, spiral, and rod-shaped, Gram-negative bacteria with an unsheathed flagellum at one or both ends. *Campylobacter* spp. are both zoonotic pathogens and the leading cause of foodborne gastroenteritis in humans worldwide.[1] In the United States, *Campylobacter* spp. cause an estimated 845,000 illnesses per year (or 9% of defined human foodborne illnesses) and are responsible for 76 deaths annually.[2] There are more than 214,000 confirmed *Campylobacter* cases in the European Union annually, at a rate of 45.2 cases per 100,000 people each year.[3] In developing countries, *Campylobacter* infection is hyperendemic among young children less than 5 years old.[4] There are currently 17 species and 6 subspecies in the *Campylobacter* genus.[5,6] *Campylobacter jejuni* and *Campylobacter coli* are the most important human pathogens in the genus, with the former usually responsible for ~90% and the latter responsible for ~10% of *Campylobacter* enteric infections.[7–9] Most other *Campylobacter* spp. also cause a range of diseases in humans and/or animals.[7]

Acute gastrointestinal disease caused by *Campylobacter* spp. is called campylobacteriosis. The main transmission route of *Campylobacter* is via ingestion of contaminated food or water. Infection of domesticated animals is widespread including poultry, pigs, sheep, cattle, dogs, cats, and birds. *Campylobacter* can be transmitted to the human population through improper handling of or consumption of undercooked poultry, pork, and beef, unpasteurized milk, contaminated drinking water, and the feces of infected pets.

The common symptoms of human campylobacteriosis include diarrhea, fever, abdominal cramps, malaise, and headache. Although generally self-limiting, approximately 10% of laboratory-confirmed cases require hospitalization. In addition to causing primary intestinal infections, *Campylobacter* can trigger serious post-infectious sequelae such as reactive arthritis (ReA) (or Reiter's syndrome), Guillain–Barré syndrome (GBS), post-infectious irritable bowel syndrome (IBS), or inflammatory bowel disease (IBD).[10] The most severe post-infectious sequela is GBS that manifests as an acute autoimmune neuropathy affecting the peripheral nervous system and leads to a typical ascending paralysis, resulting in temporary or permanent physical disability. Figure 3.1 shows major human body systems affected by *C. jejuni*.

This chapter will present an overview of the general concepts, mechanisms, and new application of analytical and molecular biological techniques for detecting, reducing, and preventing *Campylobacter* contamination of food as well as for diagnosis and treatment of infected individuals. We will focus on *C. jejuni* and *C. coli*, which are the main causes of *Campylobacter* enteritis in humans.

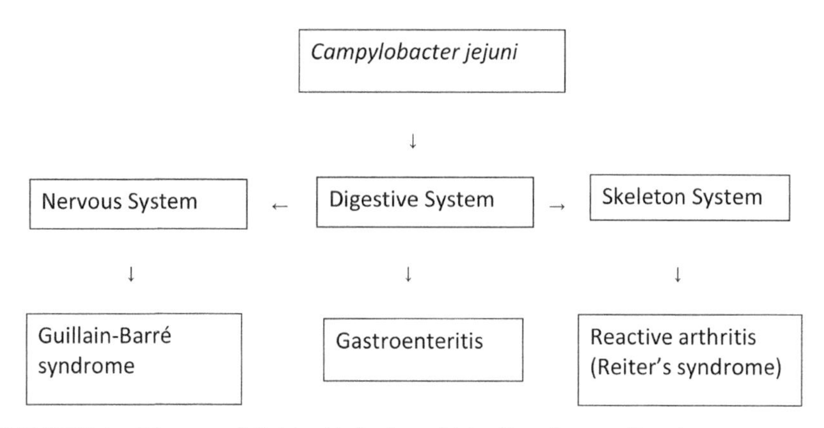

FIGURE 3.1 Diagram of *C. jejuni* infection which affects human digestive, nervous, and skeleton systems.

3.2 BIOLOGICAL CHARACTERISTICS

Campylobacter are small, non-spore-forming, spiral Gram-negative rods. The size of a *Campylobacter* bacterium is 0.2–0.9 μm in width and

0.5–5.0 µm in length. The motility of the bacteria is characteristically rapid and darting in corkscrew fashion when observed by phase-contrast microscopy. *Campylobacter* are typically microaerophilic, which means that they require limited oxygen, but are killed by normal oxygen levels in the atmosphere (approx. 20%). The oxygen-sensitive nature results in the difficult diagnosis of *Campylobacter* infection. For optimal growth, *Campylobacter* require 3–15% O_2, 3–5% CO_2 and 85% N_2, but they have also been found to grow anaerobically. All *Campylobacter* grow at 37°C, but thermophilic *Campylobacter* including *C. jejuni*, *C. coli* and *C. lari* can grow at 42°C. *Campylobacter* do not ferment or oxidize carbohydrates, do not hydrolyzed gelatin, and do not need blood components for growth. The thermophilic campylobacters reduce selenite, are oxidase and catalase positive, and are indole-negative. They can be differentiated on the basis of some biochemical ability such as nalidixic acid sensitivity and hippurate hydrolysis.[11]

Campylobacter can survive in water at 4°C for many weeks, but typically survive only a few days at temperatures above 15°C. Although freezing and thawing causes a 1–2 $\log_{10}$ fall in viable organism numbers, they can survive for 2–5 months at −20°C. *Campylobacter* are sensitive to acidity lower than pH 5. *Campylobacter* can survive in 2% or higher concentrations of sodium chloride for several weeks at 4°C, so they can survive in salted, uncooked meat if the initial level of contamination is high. *Campylobacter* are more sensitive to stress conditions such as drying, heat, freezing, acidity, disinfectants, and radiation than most other enteric pathogenic bacteria.[12] This suggests that they are best adapted for existence *in vivo*, not *in vitro*. However, *C. jejuni* produces biofilms, which may contribute to its survival under aerobic conditions and its dissemination in food processing environments.[13]

During prolonged cell culture and under stress conditions, *Campylobacter* cells gradually transform to coccoidal forms and become increasingly difficult to culture.[14] They may also enter a viable, but non-culturable (VBNC) form.[15,16] Oxygen-quenching agents in growth media, such as hemin and charcoal as well as a microaerobic atmosphere and preenrichment, can significantly improve *Campylobacter* recovery.[17]

3.3 EPIDEMIOLOGY

3.3.1 INCIDENCE

Campylobacter spp. are the leading cause of foodborne gastroenteritis in humans worldwide. The incidence of campylobacteriosis has risen recently in the United States, with 43,698 diagnosed cases and an estimated 845,024 total cases annually[2]. These bacterial pathogens cause 9.4% of cases of foodborne illnesses each year,[2] and the estimated annual cost of the illness is $1.7 billion.[18] The number of *Campylobacter* outbreaks is on the rise, since 15 outbreaks were reported in 2009, 25 in 2010, 30 in 2011, and 37 in 2012.[19]

In Europe, campylobacteriosis is also the most common foodborne bacterial illness. More than 214,268 confirmed cases were reported in the European Union in 2012,[3] with an estimated annual incidence rate of 9.2 million and an annual cost of E2.4 billion.[20] In New Zealand, campylobacteriosis was reported at 4 per 1000 people in 2012.[6]

In developing countries including Bangladesh, Indonesia, India, Gambia, and Mexico, *C. jejuni* infection occur frequently in young children.[4] The estimated campylobacteriosis occurrence in community-based studies is 60% in children under 5 years of age.[21] *Campylobacter* enteritis in developing countries intends to be watery diarrhea infected simultaneously with multiple pathogens and multiple *Campylobacter* strains.[4]

3.3.2 INFECTIOUS ROUTES

Campylobacter spp. are typically unable to multiply outside of a host, but they can survive in different environmental sources.[8] The survival time is dependent upon the species, temperature, oxygen levels and nutrition of the environments.[22]

Major reservoirs for *Campylobacter* are warm-blooded farm animals (i.e., poultry, pigs, cattle, and sheep), wildlife, and domestic pets. Animal carriers are generally asymptomatic, but *C. fetus* can cause diarrhea and aseptic abortions in large animals.[23] Handling and consumption of raw or undercooked poultry products have been determined to cause up to 30% of human transmission.[19] For example, *Campylobacter* were found on 47% of raw chicken samples brought in grocery stores in 2011[24] The prevalence of *Campylobacter* in broiler flocks and the risk to cause gastroenteritis are closely related.[25] By reducing the number of live *Campylobacter* in broiler intestines, the risk to cause gastroenteritis is dramatically reduced.[20] Both *C. jejuni* and *C. coli* easily colonize in the intestines of domestic poultry. Eating contaminated, undercooked pork or sausages can result in infection. Raw, or contaminated, poorly cooked fish, shellfish, and mushrooms can also lead to infection. *Campylobacter* do not typically multiply in foods left at ambient temperatures due to their microaerophilic nature, so they generally do not cause direct food poisoning. However, their relative low infectious dose (as low as 50 bacteria) allows the organisms to exhibit high infection rates.[26,27]

Although a major source of human infection is contaminated meat, there is evidence that surface and/or drinking water are also important sources for *Campylobacter* transmissions to humans.[28] Natural water surfaces are frequently contaminated with thermotolerant *C. jejuni*, *C. coli*, and *C. lari* throughout the year by raw sewage and feces from wild or domestic animals and discharge of wastewater from agricultural lands.[29,30] Infection can result from drinking unchlorinated, contaminated water, or the consumption of food prepared with untreated/improperly treated water. The first reported waterborne campylobacteriosis outbreak in the United States occurred in Burlington, Vermont, in which a contaminated community water supply affected an estimated 3000 people.[31] In fact, *Campylobacter* were unappreciated as important human pathogens until this outbreak in 1982.

Raw and unpasteurized milk is also a common source of *Campylobacter* and accounted for about half of the foodborne *Campylobacter* outbreaks in the United States and in Europe in 1970s and 1980s. A school field trip to a dairy farm where drinking raw milk was part of the experience serves as a typical example.[32] Although outbreaks due to raw milk have markedly decreased with routine milk pasteurization in developed countries and increased public awareness, consumption of raw (unpasteurized) milk is still associated with campylobacteriosis.[33,34] Other less common sources of *Campylobacter* are sewage, raw vegetables and fruits, as well as insects.[35]

3.3.3 INFECTIOUS DOSE

The infectious dose of *C. jejuni* is low and similar to that of *Shigella* spp. Volunteer studies have established that as few as 50–500 organisms are sufficient to initiate typical disease, but an increased infectious dose of

Campylobacter does not correlate with increased disease severity.[26] Milk, food, or medical therapies that neutralize/reduce gastric acid are thought to effectively decrease the infectious dose.

3.3.4 POPULATION FACTORS

Susceptible populations. In developed countries, *Campylobacter* affect all age groups. Infants and young children have the highest rates of infection, and young adulthood (15–29 years of age) has the second highest rates of infection.[36] Surprisingly, a recent study showed that older populations (≥ 60 years of age) are at the highest risk for campylobacteriosis in England and Wales.[37] Campylobacteriosis is hyperendemic in developing countries. Children from an early age are repeatedly exposed to *Campylobacter* and develop substantial immunity that persists into adulthood. *Campylobacter* infections are usually asymptomatic in adults and older children,[38] and occur more frequently in males than in females.[39]

Peak seasons. The incidence of campylobacteriosis shows a seasonal change with peaks in the summer and early fall in most developed countries.[40] In the United States, there is also an increased incidence in December.[36] *Campylobacter* infection is most prevalent in spring in Australia.[41] This somewhat disparate pattern correlates with varied seasonal consumption of food at barbecues and catered social events in these different countries.

Trends that increase the occurrence of foodborne illness. Large-scale production and globalization of the food supply both increase the possibility of extensive spread of foodborne diseases. More people attend outside picnics during warm weather, and more social events increase the likelihood of greater outbreaks of foodborne bacterial pathogens. Additionally, increasing levels of antibiotic-resistance and genomically evolving *Campylobacter* may enhance the organisms' potential to spread.[42]

3.4 DISEASES

3.4.1 ENTERITIS

The average incubation period of campylobacteriosis is about 3–7 days from ingestion of *Campylobacter*-contaminated food or water, or occupational exposure to *Campylobacter*-infected animals. The illness may start from abdominal cramps and watery diarrhea, but reverts to fever and small volume stools that contain blood and leukocytes with a frequency as high as 10 times per day. Transient bacteremia also occurs. Patients may experience fever as high as 40°C, chills, headache, malaise, and occasionally delirium. Nausea is common, but vomiting is rare. The acute infection may last 2–10 days and can be treated with appropriate antibiotics. However, antibiotic-resistance is a growing concern with *Campylobacter*.

As it is common with many infectious diseases, up to one quarter of exposed individuals have no noticeable symptoms. In most patients, the illness is self-limiting, and recovery occurs within one week. In developed countries, fecal leukocytes are found in about 75% of patients, and gross or occult blood is seen in approximate 50% of patients. *Campylobacter* tend to cause a non-inflammatory, watery diarrhea. Untreated patients may carry and shed bacteria from one month to several months.[27]

The clinical symptoms of *Campylobacter* enteritis are similar to those of other bacterial enteritis, such as *Salmonella* or *Shigella* enteritis. However, abdominal pain tends to be more severe in *Campylobacter* infection. Sometimes, the abdominal pain appears to be more significant than the diarrhea, mimicking acute appendicitis (i.e., pseudoappendicitis) or pancreatitis.

3.4.2 ACUTE COMPLICATIONS

Bacteremia. *C. jejuni* bacteriemia occurs in about 0.15% of enteritis,[43] which is higher in elderly individuals and in the immunocompromised, and is higher in males than females. Although transient bacteremia might occur in *C. jejuni* enteritis at a higher rate, it is difficult to be diagnosed because blood cultures are not routinely acquired. Serious bacteremia does not occur very often, since most *C. jejuni* strains are killed by normal human serum. In contrast, *C. fetus* is serum-resistant because it is covered with an S layer, a capsule-like protein that prevents complement-mediated killing. *C. fetus* infections frequently induce bacteremia and systemic spread. *C. upsaliensis* and *C. hyointestinalis* also cause bacteremia. However, *C. jejuni* still accounts for 90% of *Campylobacter* isolated from human blood samples.[27]

Other infections. Secondary acute complications resulting from primary *Campylobacter* enteritis include cholecysitis, pancreatitis, peridontitis, gastrointestinal hemorrhage, meningitis, endocarditis, arthritis, peritonitis, cellulitis, hepatitis, and septic abortion. These complications occur infrequently. Recently, a typical case of acute pancreatitis was reported to be associated with *Campylobacter* enteritis.[44] More *Campylobacter* species are identified in human infections, especially in immunocompromised patients, with improving bacterial identification methods. *C. fetus* is reported to cause human fetal death or septicemia. Both *C. jejuni* and *C. coli* cause human abortion.[45] *C. concisus*, *C. rectus*, and *C. curvus* have been implicated in human periodontal disease and other diseases.[46]

3.4.3 CHRONIC COMPLICATIONS

A significant proportion of *Campylobacter* patients develop serious chronic complications. Recently, a systematic review and meta-analysis estimated that the proportion was 2.86% for developing ReA, 4.01% for developing post-infectious IBS, and 0.07 % for developing GBS after *Campylobacter* infection.[47] These chronic sequelae, following campylobacteriosis, cause significant social, medical, and economic burdens.[48]

Reactive arthritis (or Reiter's syndrome). This syndrome encompasses a reactive arthropathy involving multiple joints and affects approximately 3% of *Campylobacter*-infected patients.[48] ReA usually develops 2–4 weeks after the infection, and occurs most often in young men aged 20–40 years. Arthritis, conjunctivitis, and urethritis are three major symptoms. Approximately 75% of patients with ReA have a positive blood test for the genetic marker HLA-B27 phenotype.[49]

Guillain–Barré Syndrome. GBS is a disorder affecting the peripheral nerve system resulting in neuromuscular paralysis, and is triggered after certain viral and bacterial infections. About 30% of all GBS cases are estimated to be triggered by *Campylobacter* infection.[50] In certain patients, respiratory failure precipitates the requirement for mechanical ventilation. Approximately 5% of patients with GBS die, and 20% of patients are left with some chronic disability.[51] Importantly, GBS is estimated to occur in approximately one individual per 1000 *C. jejuni* infections.[52] The pathogenesis of GBS is believed to involve molecular mimicry between *C. jejuni* surface polysaccharides and certain peripheral neural gangliosides (such as GM_1, GD_{1a}, GD_3, and GT_{1a}).[53,54] *C. jejuni* Penner serotypes O:19 and O:41 are most commonly associated with GBS.[55,56]

Post-infectious irritable bowel syndrome. IBS is an inflammatory chronic bowel disease, which causes abdominal pain, cramping, bloating, gas, and diarrhea or constipation.

Inflammatory bowel disease. The idiopathic IBDs comprise Crohn's disease and ulcerative colitis, which may result from an inappropriate inflammatory response to intestinal microbes in a genetically susceptible host. In contrast with that of Crohn's disease, the inflammation of ulcerative colitis is limited to the colonic mucosa. Infection with *Campylobacter* is associated with an 8–10 times higher risk of developing ulcerative colitis in the following year. The risk diminishes with time, but is still present after 10 years.[57,58]

3.5 GENOMICS

Genomic analysis is an important tool for understanding bacterial biology and virulence. *C. jejuni*, *C. coli*, and *C. fetus* have genomes of approximately 1700 kb, which is only about one-third the size of the well-studied *Escherichia coli* bacterial genome. This small genome size for campylobacters may relate to their inability to ferment carbohydrates and to degrade complex substances, as well as their requirement for special growth conditions.[59]

The first available *Campylobacter* DNA sequence data were obtained from the clinical isolate, *C. jejuni* NCTC 11168.[60] This isolate has a circular, highly dense, and rather small genome (~1.6 Mb) with a 30% G+C content. Surprisingly, this *C. jejuni* genome lacks any insertion elements or prophages. Re-annotation and re-analysis of this *C. jejuni* genome revealed that the complete genome is 1,641,481 bp in length with 25 polymorphic regions.[61] The re-annotation removed 11 coding sequences from the total original number of 1655, added additional information to 90% of the coding sequences, and added function to 18.2% of the coding sequences.[61] One unique feature of *Campylobacter* is the common appearance of homo-polymeric nucleotide stretches within the genome that, during replication, lead to slip-strand repair mutagenesis at high frequency. These high frequent mutations make genetic and pathogenicity studies of *Campylobacter* quite challenging, due to a constantly changing genetic landscape.

3.5.1 FLAGELLA

The flagellum is composed of the major subunit FlaA and the minor subunit FlaB, encoded by genes *flaA* and *flaB*, respectively.[62] The polar flagellum and unique characteristics of the spiral shape of campylobacters confer a distinctive darting motility, that is, particularly effective in a viscous matrix. *Campylobacter* flagella have been implicated as adhesins.[63] Nonflagellated mutants are impaired in adherence to INT407 cells, which can be enhanced by bacterial centrifugation onto the monolayer.[64] Motility is required for optimal *Campylobacter* adherence to host cells and for invasion ability.

3.5.2 TOXINS

There are two reported types of *Campylobacter* toxins: enterotoxin and cytotoxins. Infected individuals do not develop neutralizing antibodies to these toxins.

Cytolethal distending toxin (CDT) is reported to be a virulence factor for *Campylobacter* pathogenesis in humans and animals. CDT is encoded by three adjacent genes, *cdtA*, *cdtB*, and *cdtC*.[65,66] These three subunits are required for full toxin activity; *cdtB* encodes the active/toxic component of the toxin, while *cdtA* and *cdtC* are essential for mediating toxin binding to and internalization into the host cell.[67] CDT is encoded by approximately 40% of *Campylobacter* strains tested. After being translocated into the nucleus, CdtB can cause double strand DNA breakage, block replication at the G2/M phase of the eukaryotic cell cycle, and ultimately cause host cell death.[68,69] CDT is thought to contribute to *Campylobacter* pathogenesis by inhibiting immunity, and causing deaths of epithelial cells and fibroblasts, which may result in enhancement of disease symptoms and in prolonged healing.[69,70]

The enterotoxin of *C. jejuni* is similar to the related *Vibrio cholerae* and *E. coli* heat labile enterotoxins (i.e., CT and LT).[60] This enterotoxin is produced to a lesser degree by *C. coli*. It has been suggested that this enterotoxin causes watery diarrhea, but enterotoxigenic strains have been isolated from asymptomatic carriers, and strains lacking apparent enterotoxic activity are still fully virulent. Thus, the role, if any, played by enterotoxin in *Campylobacter* pathogenesis is yet undefined.

There are several cytotoxins, aside from CDT, identified in *Campylobacter* spp. They consist of a 70 kDa cytotoxin, a Vero/Hela cell cytotoxin, a shiga-like toxin, a hemolytic cytotoxin and a hepatotoxin.[71] The roles of these reported "cytotoxins" in *Campylobacter* pathogenesis remain uncharacterized.

3.5.3 LIPOOLIGOSACCHARIDES (LOS)

LOS are found on the surface of many mucosal bacterial pathogens. *Campylobacter* species express a LOS (i.e., LOS core linked to→lipid A) and a capsular polysaccharide (CPS) instead of a full length LPS (O-chain →LOS core→ lipid).[72–74] *C. jejuni* LOS is important in host cell adhesion, invasion into human intestinal cells, and protection from complement-involved killing.[63]

Godschalk et al.[75] determined that specific LOS classes contain sialylation genes which are associated with GBS. Molecular mimicry between bacterial sialylated LOS structures and human neural ganglioside epitopes (e.g., GM_1, GD_{1a}, GD_3, and GT_{1a}) that generates cross-reactive immune response results in autoimmune-driven nerve damage.[76–78] Immunization studies in rabbits and mice have demonstrated that these anti-ganglioside antibodies can be induced using purified LOS from *C. jejuni*, but require a strong adjuvant.[79] Importantly, yet undefined, host-dependent factors are thought to be important in the generation of anti-ganglioside antibodies following *C. jejuni* oral infection of chickens.[80]

3.5.4 CAPSULAR POLYSACCHARIDE (CPS)

The CPS of *C. jejuni* is the major serodeterminant of the Penner serotyping scheme.[73] CPS contributes to serum resistance, invasion of intestinal epithelial cells, and virulence in the ferret and other animal models.[81,82] There are 47 serotypes of *C. jejuni*, 22 of which fall into complexes of related serotypes.[83] Besides variation in sugar composition, the CPS can be modified with ethanolamine, glycerol, and O-methyl phosphoramidite (MeOPN). The MeOPN modification has been found in about 70% of *C. jejuni* strains sequenced to date. A CPS conjugate vaccine from strain 81–176 shows significant protection against diarrheal disease in a nonhuman primate model of diarrhea.[84] The capsule polysaccharide of *C. jejuni* and the MeOPN modification modulate the host immune response.[85]

C. jejuni glycan surface structures are important for both host cell adherence and invasion processes. Inactivation of the capsule biosynthesis genes *kpsM*[81] and *kpsE,*[86] the LOS biosynthesis gene *gale,*[87] and the genes *pglB* and *pglE*[88] cause an adherence/invasion defective phenotype. Stahl et al.[89] demonstrates in a mouse model that *C. jejuni*'s capsule could inhibit the host's immune response, since capsule loss led to significantly increased activation of host TLRs and exaggerated gastroenteritis.

3.5.5 PROTEIN SECRETION SYSTEMS

Gram-negative bacteria have evolved at least six different protein secretion systems to export proteins to the periplasmic space, into host cells, or to the environment. It is reported that *C. jejuni* display Type III, Type IV, and Type VI secretion systems.[90–93]

Certain *Campylobacter* proteins (e.g., *Campylobacter* invasion antigens (Cia), are exported from the bacterium's flagellar T3SS and are delivered to the host cell where they may play a role in promoting cell invasion.[93] Only three of 18 Cia proteins (CiaB, CiaC, and CiaI) have been identified in host cells.[92,93] CiaI is suggested to be involved in efficient intracellular survival.[92] However, it should be emphasized that the involvement of Cia proteins in host cell invasion is controversial.[94]

The Type VI secretion system (T6SS) forms a needle-like structure that transports toxic effector molecules into neighboring eukaryotic cells or bacteria. *C. jejuni* was reported to contain a functional T6SS gene cluster.[91] The T6SS gene cluster contains 13 conserved genes, encodes tail subunit-like, hemolysin co-regulated proteins (Hcp) and valine glycine repeat G (VgrG) proteins, and is present in ~10% of *C. jejuni* strains. Hcp may polymerize into the T6SS needle structure, and VgrG proteins may form the spike of the nanotube in the secretion system. *C. jejuni* T6SS causes contact-dependent lysis of red blood cells, which is only observed in capsule-deficient bacteria.[91]

3.5.6 ADHERENCE/INVASION RELATED GENES

Numerous genes have been reported to be involved in *Campylobacter*'s ability to adhere to and invade host cells. SodB, an active superoxide dismutase has been reported to somehow be involved in efficient adherence and invasion, but its most import role may be in intracellular persistence.[95] A fibronectin-like protein A (FlpA),[96] an autotransporter protein CapA,[97] and a putative lipoprotein (Cj0497)[98] have all been identified as adherence factors.

It has also been reported that PEB1 (a 28 KDa surface antigen), JlpA (a 42.3 KDa surface-exposed lipoprotein), CadF (a 37 KDa outer membrane protein), and PglH (N-linked general protein glycosylation pathway) are involved in *C. jejuni* adherence to host cells.[99–102] Peb1A has been characterized as a periplasmic aspartate-glutamate binding protein required for the utilization of certain amino acids as growth substrates in vitro, and *peb1A* mutants showed no significant invasion defects.[103,104] Peb1A may play a crucial role in the proliferation and persistence of *C. jejuni in vivo* as demonstrated in various animal infection studies.[96,105]

3.5.7 BIOFILM FORMATION

Biofilm formation increases *Campylobacter* survival in harsh environment.[106,107] Although the genes of *C. jejuni* biofilm formation are not well understood, several studies of targeted deletions in *C. jejuni* strains 81–176 and 11168 have shed lights in the field.[106–109] Table 3.1 summarizes major genes involved in biofilm formation in *C. jejuni.*

TABLE 3.1 The Reported Genes Involved in Biofilm Formation in *Campylobacter* spp.

Gene	Encoded protein	Function	Reference
cadF	Fibronectin-binding protein	Adhesion, biofilm formation	[108]
capA	*Campylobacter* adhesion protein A	Adhesion, invasion, biofilm formation	[96, 97]
cheA, cheV,cheW, cheY	Chemotaxis protein A/V/W/Y	Chemotaxis, biofilm formation	[108]
cprRS	*Campylobacter* planktonic growth regulation	Regulation of biofilm formation	[146]
csrA	Carbon starvation regulator	Regulation of motility, adherence	[147]

TABLE 3.1 *(Continued)*

flaAB/flaC/flaG	Flagellar structure protein/flagellin subunit protein C/G	Motility, adherence, regulation of biofilm formation	[108, 109, 148]
flgE/flgG/flgG2,	Flagellar hook protein/ putative basal body protein G/G2	Motility, biofilm formation	[108]
fliA/fliV/fliD/fliS	Class III flagellar sigma factor A/V/flagella filament cap/flagella protein	Biofilm formation and regulation	[108]
flgK	Putative hook-associated protein	Biofilm formation	[108]
flhA	A protein of flagella export apparatus	Biofilm formation	[108]
groES/groEL	Co-chaperonin/ chaperonin	Biofilm formation	[108]
kpsM	CPS export protein	Adherence, invasion, biofilm formation	[106]
lgtF	prolipoproteindiacylglyceryl transferase, LOS structure	biofilm formation	[149]
luxS	Quorum sensing	Biofilm formation	[109]
peb4	Cell binding factor	Adhesion, biofilm formation	[108]
ppk1/ppk2	Polyphosphate kinase 1/2	Adhesion, biofilm formation	[150, 151]
phoX	Poly P associated protein	Biofilm formation	[151]
pseA	Flagellar glycosylation protein	Adhesion, biofilm formation	[152]
racR	Two-component regulation	Biofilm formation	[108]
tatA/tatB/tatC	Twin-arginine translocase A/B/secretion system	Biofilm formation	[153]

waaC/waaD/waaF	Heptosyltransferase I/ADP-L-glycero-D-mannoheptose-6-epimerase/LOS structure	Biofilm formation and virulence factors	149, 154

3.6 PATHOGENESIS

Bacterial diseases of the gastrointestinal tract typically result from a complex set of interactions between the offending bacteria and the host. The pathogenesis of *Campylobacter* gastrointestinal disease is not completely understood. The results of intestinal biopsies of patients, infected primates, and several other experimental model animals, as well as cell culture assays have demonstrated that *C. jejuni* is able to invade enterocytes and suggest that gut adherence, invasion, mucosal translocation, eliciting host immune and inflammatory responses are important steps in pathogenesis.

After ingestion via contaminated food or water, *Campylobacter* have to pass through the gastric acid barrier of the stomach, and the highly alkaline secretions from the bile duct in the upper small intestine. Reduction of gastric acidity by antacid use or other medicines increases the risk of *Campylobacter* infection. The organisms apparently first colonize the jejunum and ileum, and then affect the colon and rectum. Mucosal damage and inflammation is observed in both the small and large bowels. In developed countries, *C. jejuni* causes an invasive, inflammatory disease; however, in developing countries, *Campylobacter* tends to cause a non-inflammatory watery diarrheal disease.[94]

3.6.1 ADHERENCE

Adhesion of *C. jejuni* to host cells plays an important role in colonization of chickens and in human infections. *Campylobacter* binding to host cell receptors is not mediated by fimbriae or pili, like some enteric pathogens. Reported adhesins include flagella, outer membrane proteins, and surface polysaccharides. Functional flagella are involved in epithelial cell adherence and are essential for the internalization of *C. jejuni* into cultured epithelial cells and for colonization of the mouse intestines.

Rubinchik et al.[110] report that PEB3 is a cell surface glycoprotein required for bacterial interaction with a host cell receptor. These investigators also demonstrated that the production of capsule reduces bacterial attachment, and that the genes involved in capsule and PEB3 biosynthesis are separately regulated.

Although some putative *C. jejuni* adhesins have been identified, the molecular mechanisms of their interaction with host cells and their exact roles in pathogenesis remain to be elucidated.

3.6.2 INVASION

Campylobacter invasion into the epithelial mucosa appears to be an essential process leading to colitis. Bacterial internalization has typically been observed to involve host cytoskeletal structures resulting in endocytosis of the pathogen. The cytoskeleton of the eukaryotic cell is a complex array of proteins, the most prominent of which are microfilaments and microtubules. Most invasive enteric organisms including *Salmonella, Shigella, Yersinia,* and *Listeria* have been found to trigger largely microfilament-dependent entry pathways. Studies of *Campylobacter* invasion suggest that the organisms may encode separate microtubule-dependent and microfilament-dependent pathways for host invasion in different cell lines or under different cell culture condition.[94,111]

Cell-culture invasion assays conducted with biochemical inhibitors have been employed to study the host cell signal transduction pathways triggered by interaction with C. *jejuni* strain 81–176. Divalent calcium cation (Ca^{2+}) plays a pivotal role in host signal transduction and other cellular processes. The essential involvement of Ca^{2+} release from host intracellular stores for *C. jejuni* 81–176 internalization into host cells has been reported.[112] Host signal transduction studies have suggested that strain 81–176 interacts at the host cell surface with G protein-coupled receptors, a process which activates PI-3 kinase and MAP kinases (p38 and ERK) that appear to be intimately involved in microtubule-dependent events controlling 81–176 internalization.[113] The rho GTPases, Rac1, and Cdc42 have also been implicated in the 81–176 invasion process.[114] In a mouse knockout cell line, fibronectin, integrin-β1, and the guanine exchange factor Vav2 appear to trigger Cdc42 activation and to induce filopodia and membrane dynamics.[115]

3.6.3 MUCOSAL TRANSLOCATION

The intestinal mucosa forms a barrier that protects against invasion of the host by nonpathogenic bacteria residing in the intestinal lumen. Bacterial penetration of enterocytes is an important step in pathogenesis. *Campylobacters* have been observed to translocate across a tight epithelial cell monolayer.[94] *C. jejuni* can penetrate from the apical to the basolateral surface of polarized Caco-2 cells without disrupting transepithelial electrical resistance.[116] Electron microscopic studies also show that *Campylobacter* pass between cells; that is, some isolates appear to transcytose paracellularly without host cell endocytosis. Kinetics analyses and EM studies of *Campylobacter*-host interaction have demonstrated unique host cell adherence, early endocytosis-specific structures, and an exocytosis component of the transcellular transcytosis route of entry.[117] Thus, it appears that *Campylobacter* may cross polarized epithelial cells via both transcellular and paracellular routes.[117] Recently, a study showed that the mucous produced by a HT29-derivative E12 cell line on transwell filters increased the binding to and invasive efficiency of *C. jejuni* for E12 cells; probiotics decreased the binding and invasive efficiency of *C. jejuni.*[118]

C. jejuni serine protease HtrA (high temperature requirement A) is presented to be a novel virulence factor.[119] HtrA is secreted into the extracellular space where it cleaves cell surface adhesion proteins and tumor-suppressor E-cadherin.[120,121] Mutants in *htrA* lack E-cadherin shedding, and could not transcytose across polarized human MKN-28 epithelial cells.[116]

3.6.4 BACTERIAL SURVIVAL AND THE IMMUNE RESPONSE

Following invasion into cells, pathogens must have the ability to survive within and/or replicate intracellularly. The bacterial and host factors that determine the fate of internalized *Campylobacters* are not well understood. Internalized *C. jejuni* change from the spiral to a coccoid form within 4–8 h and survive for 6–7 days in human mononuclear phagocytes.[122] Monocytes, therefore, could play a key role in dissemination of *C. jejuni* following intestinal translocation by protecting *C. jejuni* during transit to secondary sites. Some studies report that *C. jejuni* can survive in monocytes or macrophages,[123,124] but other studies suggest that *Campylobacter* are effectively killed by macrophages.[71,125] Further investigation is required to clarify this important issue.

Host defense against microbial pathogens involves coordination of multiple signals between cells of both the innate and acquired immune systems. Interleukin-8 (IL-8) secretion by epithelial cells may be an early signal for the acute inflammatory response following various enteric bacterial infections. Many strains of *Campylobacter* spp. can induce secretion of IL-8 by INT407 cells.[126] *Campylobacter* also trigger an innate inflammatory response through increased production of IL-1β, IL-6, IL-8, IL-12, and TNF-α and initiate a Th1-polarized adaptive immune response in human epithelial cells and monocyte-derived dendritic cells (DCs).[127] Also, selected proinflammatory chemokines are induced after infection of human epithelial cells and DCs with *C. jejuni.*[128]

Sialic-acid binding Ig-like lectins (Siglecs) have emerged as important players in host immunity.[129] Siglec10 is an immune-modulatory receptor. Viable *C. jejuni* and purified flagellum can bind Siglec10. Infection of Siglec 10 overexpressing DCs results in increased IL-10 expression in a p38-depecdent manner in vitro. *C. jejuni* flagella promote an anti-inflammatory axis via glycan-Siglec 10 engagement.[130]

3.7 IDENTIFICATION AND DETECTION

Robust and accurate detection assays are needed to detect causative infectious agents rapidly in order to choose appropriate treatment and to limit their spread. Many bacterial culture methods, molecular methods, and immune techniques have been developed for detection and identification of *Campylobacter* spp. However, bacterial culture, which may take 2–6 days, is required for accurate *Campylobacter* diagnosis. Following detection by culture, more refined methods can be used for typing and determination of antimicrobial susceptibility.

Molecular methods, which identify the source of a patient's or an environmental *Campylobacter* isolate by its genome has been used in many epidemiological studies and clinical practices.[131,132] A number of molecular methods such as pulsed-field gel electrophoresis (PFGE), multilocus sequence typing (MLST) scheme, polymerase chain reaction (PCR) have emerged as the leading method for *Campylobacter* identification and detection.[133]

3.7.1 DIRECT EXAMINATION

It is possible to detect *Campylobacter* infection by observing directly the characteristic Gram-negative nature, curved or spiral morphology and the characteristic darting motility in wet mounts of fresh stools detected by phase-contrast microscopy or dark field microscopy.

3.7.2 TRADITIONAL CULTURE METHOD USING SELECTIVE AGAR

Campylobacter multiply more slowly than the other enteric microbial flora and require low oxygen levels; and therefore cannot be isolated from fecal specimens without using selective media. The most common selective agar used is charcoal cefoperazone deoxycholate agar (CCDA). Plates are incubated at 37°C for 2 days in anaerobic jars or in a Campy gas atmosphere. The colonies of *Campylobacter* usually are gray, flat, irregular, and spreading in freshly prepared media. Selective culture is a quick, cheap, and effective method for identifying *C. jejuni* and *C. coli* from fecal samples.[134] Humans infected with *Campylobacter* usually excrete 10^6–10^9 *C. jejuni* colony-forming unit per gram stool; thus, enrichment is usually unnecessary for stool samples. The drawbacks include that agar plates maybe overgrown by faster growing microorganisms and this method does not identify the less common *Campylobacter* species.

An enrichment step is not typical necessary for clinical specimens. However, enrichment methods are necessary for food, environmental samples, and old stool samples where the number of organisms is low (sometimes, due to delayed transport of stool to the laboratory or the acute stage of diseases has passed). Organisms that have been sublethally injured by heating, freezing, or chilling should be preincubated for 4–6 h in nonselective broth at 37°C, then transferred to enrichment broth at 42°C for 44–48 h, and then inoculated from this enrichment broth to selective agar plates and incubated for a further 48 h. There is no need to add oxygen quenching substances (if grown in a special atmosphere) or blood to enrichment broth for the isolation of *Campylobacter* spp. However, the addition of blood to agar may aid in differential identification of presumptive colonies.

Antimicrobial drug sensitivity testing can be determined by using the disk diffusion method for erythromycin, tetracycline, ciprofloxacin, and other antibiotics. Molecular determination of *Campylobacter* resistance via DNA sequencing or PCR-based methods has also been performed. High levels of resistance to tetracycline and ciprofloxacin are frequently reported by many national surveillance programs, but resistance to erythromycin and gentamicin in *C. jejuni* remains low.[24]

3.7.3 MEMBRANE FILTRATION

Filtering a sample through a cellulose triacetate membrane (with 0.45–0.65 μm pores) onto an agar plate permits isolation without using antibiotic containing media. *Campylobacter* are very small and thin, so they can pass through the filters easily. The use of filtration techniques and nonselective media cultured at 37°C improves the growth of both *C. jejuni* and the atypical *Campylobacters* from stool samples, and is now recommended for primary isolation of *Campylobacter* from fecal specimens or swabs.[134]

Water analysis laboratories often filter water samples, followed by bacterial enrichment and cultivation on selective agar plates.[135] Both membrane filtration and selective culture are less efficient when applied to isolating *Campylobacter* spp. from water samples because the bacteria in the water samples may enter a VBNC state.[136]

3.7.4 TYPING

Identification of source contamination and infectious trends is vital, prior to initiating informed control measures, and this process requires subtyping of bacterial isolates. A variety of typing methods are available for *Campylobacter* spp. including the most widely used Penner serotyping scheme, MLST, and PFGE.[133]

Polysaccharide capsule typing. The Penner scheme is a passive slide hemagglutination assay that includes 47 serotypes.[72] The CPS is highly variable among strains and is a major sero-determinant of the Penner scheme.[73] However, in some serotypes LOS have been shown to play a role in Penner serospecificity.[73,83] Penner serotyping method is very complex and expensive to produce the antisera specific to all *C. jejuni* serotypes. A recently developed multiplex PCR method for determination of capsule type offers the potential of a more rapid and affordable method. The current CPS multiplex is composed of a total 14 primer pairs.[137] This multiplex system is sensitive, specific, and independent of phase variation of capsule expression in typing 244 strains.[137]

Pulsed-field gel electrophoresis. PFGE is a rapid molecular subtyping method, which is based upon the variable migration of large DNA fragment restriction fragments in an electrical field of alternating polarity. PFGE can compares the DNA fragment fingerprints of any two bacterial strains, and is used in epidemiological studies worldwide. The standard PFGE protocol for the subtyping of *Campylobacter* spp. is based on the procedure employed by PulseNet (http://www.cdc.gov/ncidod/dbmd/pulsenet/pulsenet.htm). PFGE is still a gold standard in investigating strain sources and studying bacterial genetic relationships although more rapid typing methods have been developed.

Multi-locus sequence typing. MLST is a typing technique for characterizing isolates of bacterial species using the sequences of internal fragments (400–500 bp) of multiple housekeeping genes. MLST is an epidemiological tool for identifying clinical, food, and environmental isolates, and for estimating the relative importance of each source of *Campylobacter* infection.[133]

Whole genome sequencing (WGS). WGS determines the complete DNA sequence of a *Campylobacter* genome at a single time, provides the most comprehensive analysis of genetic variation, and can differentiate down to a single base pair. However, WGS is relatively expensive and difficult to manage large databases of information, and, thus, cannot be widely employed for extensive epidemiological investigation.[131]

3.7.5 PCR

PCR is a rapid technique to amplify a short sequence of DNA in samples containing only minute quantities of DNA or RNA in a few hours. The technique offers both sensitivity and specificity in detecting bacterial pathogens.[133,138] Multiplex PCR techniques simultaneously amplify several DNA sequences and identify isolates to the species level. Real-time PCR allows for the identification of *Campylobacter* to the species level, performs absolute or relative quantification of nucleic acid copies obtained by PCR, and gets results in less than one day.[139] However, it is relative expensive and it does not identify bacterial viability.

Ethidium or propidium monoazide pretreatment of samples for qPCR (EMA- or PMA-qPCR) could solve the problem. The method pretreats a sample with intercalating dyes like EMA or PMA. These dyes can cross the membranes of damaged cells, prevent PCR amplification of DNA from nonviable cells, and ensure that only whole intact cells are amplified and enumerated. EMA- or PMA-qPCR is a rapid, quantitative detection method for viable *Campylobacter* cells from water and food samples.[140] Primer pairs utilized in PCR detection techniques for *Campylobacter* spp. include those for the identification of structural genes or potential virulence genes, such as *aspA, aspB, cdtA, cdtB, cdtC, cadF, ceuE, ciaB, flaA, flaB, glyA, hipO, iamA, pepT, pldA, virB11, ERIC,* and others.[65,90,95,139]

3.7.6 MATRIX-ASSISTED LASER DESORPTION IONIZATION TIME-OF-FLIGHT MASS SPECTROMETRY (MALDI-TOF MS)

Most new diagnostic methods for detecting *Campylobacter* spp. are laboratory-based and expensive, requiring complex equipment and specialized skills. However, rapid and cheap field-based assays have also been developed. MALDI-TOF MS provides a rapid, accurate, low-cost, and reliable identification method.[141]

Proteins and peptides can be characterized by high-pressure liquid chromatography (HPLC) or SDS PAGE by generating peptide maps. The peptide maps have been used as strain ID protein fingerprints. Mass spectrometry generates a peptide map after proteins are digested with specific enzymes. The peptide map can be used to search a sequence database to form a good match from the existing database. MALDI-TOF MS is a good tool for screening peptide masses of tryptic digests, requires relatively less intense sample preparation, and can provide an identification of a microorganism's genus and species in just a few minutes.

3.7.7 IMMUNOASSAY METHODS

Immunological techniques, both specific and sensitive to bacterial species identification, have distinct advantages in the isolation process. Bacterial culture is commonly utilized for the *Campylobacter* diagnosis. However, this diagnostic process requires at least 4 days, which is problematic in identifying life-threatening cases of campylobacteriosis, especially in infants or the elderly. Although several PCR assays have been developed, they require the use of electrophoresis to detect amplified products. Recently, commercial immunochromatography kits [including ProSpecT *Campylobacter* Microplate (Remel, Lenexa, KS), Premier CAMPY assay (Meridian Bioscience Inc., Cincinnati, OH), ImmunoCard STAT!CAMPY (Meridian Biosciences, Cincinnati, OH), Xpect *Campylobacter* assay (Remel), and RIDASCREEN (Biopharm, Darmstadt, Germany)] have become available for rapidly detecting *C. jejuni* and *C. coli* in diarrheal stool.[142]

3.7.8 BIOSENSORS

A biosensor is a device composed by a transducer and biological element. The biological element may be an enzyme, an antibody, or a nucleic acid response, which is converted into an electrical signal by the transducer. A variety of substances may be used as the bioelements in biosensors. Although there are no commercial biosensors available to detect *Campylobacter* spp. in food or other samples, biosensors could potentially serve as a powerful tool to detect *Campylobacter* spp. with the advantages of high sensitivity and specificity, real time, and on-site monitor.[142,143]

3.8 TREATMENT AND PREVENTION

3.8.1 TREATMENT

Campylobacteriosis is a self-limiting disease; and antimicrobial therapy is not generally indicated in most cases. However, if treatment is initiated early, antibiotics can reduce symptoms, shorten the period of acute illness, and control transmission within the community.[4]

Antimicrobial treatment (e.g., macrolides and fluoroquinolones) is necessary in severe, prolonged cases, or in infections in people with weakened immune systems by other chronic illnesses or medications. Antibiotic susceptibility testing/epidemiological monitoring of resistance continues to play a critical role in guiding appropriate therapy. Macrolides, quinolones, and tetracyclines are among the common antimicrobials recommended for treatment. Severe (with accompanying fevers, blood in stools) or prolonged cases may require ciprofloxacin and azithromycin. About 90% of cases respond to ciprofloxacin treatment. Fluid and electrolyte replacement may be required for serious cases of gastroenteritis.

3.8.2 PREVENTION

The annual estimated number of cases of campylobacteriosis worldwide is 400–500 million.[144] Prevention of *Campylobacter* infections is a critically important public health issue, requiring a multipronged approach, including rapid detection and treatment of diseased individuals, interruption of transmission routes, routine food inspections, and development of efficient and safe vaccines.

The transmission of *Campylobacter* to humans is attributed to the consumption of contaminated food and water, particularly poultry meat. Reducing the load of *Campylobacter* present in poultry at the farm and on the carcass is a key prevention strategy. Using bacterial phages to control the levels of *Campylobacter* in the farm and food processing environments has been successfully employed experimentally. The method is also a good tool to reduce antibiotic-resistant strains. However, phage-based products against *Campylobacter* have not yet reached the market place.[145]

Important strategies for controlling *Campylobacter* include the use of good agricultural practices, good manufacturing practices, and standard operating procedures for sanitation on the farm and within the food processing industry. Food-related workers should be trained to avoid food contamination by practicing good personal hygiene, properly cleaning and sanitizing of equipments, and preventing cross-contamination.

In the home, consumers can prevent *Campylobacter* infections by cooking all raw meat products thoroughly, washing hands with soap and running water before and after handling raw foods of animal origin, and by following food preparation recommendations. Do not drink unpasteurized milk or untreated surface water.

3.9 ADVANCES IN *CAMPYLOBACTER* RESEARCH

In recent years many advances have been made in understanding the genetics and pathogenesis of *Campylobacter* spp., and also in their diagnosis and treatment. However, we still have a long way to go to understand thoroughly the pathogenic mechanisms of *Campylobacter* infections and to develop methods for controlling them efficiently. *Campylobacter* have unique growth features, a high incidence of infection, sparsely incomplete understood pathogenic mechanisms, and no available vaccines, as well as the continual emergence of antibiotic resistant strains, posing a serious public health concern. The difficulties will require both more research and new practical applications to better control *Campylobacter* diseases.

3.9.1 DEVELOPING VALID VACCINE

To date, no effective *Campylobacter* vaccine is available. Flagellin, CPS, and some virulence gene mutants have been reported as good antigenic candidates for *Campylobacter* vaccine developments. However, no safe and effective *Campylobacter* vaccine has yet been developed.

3.9.2 FINDING RELEVANT AND CONVENIENT ANIMAL MODELS

*Campylobacte*r remain poorly characterized in terms of pathogenic mechanisms, largely because of absence of small animal models that mimic human diseases. Although lacking typical disease symptoms following infection, some animal models such as ferrets and mice have been useful in examining colonization and immune response.[84,89] The use of animal models is limited by the difficulty of obtaining and maintaining relevant and convenient animals and lack of sound evidence of the relationship between the inflammatory and immunologic reactions of these animals and those of human infections. Thus, obtaining and studying relevant and convenient animal models is challenging, but necessary to understand *Campylobacter* pathogenesis and for future to evaluation of candidate vaccines.

3.9.3 DEVELOPING RAPID AND EFFECTIVE DETECTION

METHODS

Improving current *Campylobacter* detection methods and developing new molecular identification approaches will aid rapid and efficient detection and identification. Low cost of these methods and easy-of-use will also be important in developing countries.

3.9.4 STUDYING ON PATHOGENESIS

Campylobacter remain a leading cause of gastroenteritis in the world. A clearer understanding of the molecular mechanism of *Campylobacter* enteritis and post-infectious sequelae is an ongoing research issue. Many important questions await further study such as: what are the major bacterial ligand(s) and host receptor(s) for *Campylobacter* uptake? How to evaluate the importance of the protein secretion systems present in *C. jejuni*? How do *Campylobacter* survive in human epithelial cells and monocytes? How do *Campylobacter* cause GBS and other post-infectious sequelae? Are the pathogenic mechanisms of *Campylobacter* enteritis similar in developed versus developing nations? What are the relevant immune responses triggered by *Campylobacter* infection that prevent from future *campylobacter* disease?

3.9.5 PREVENTING CAMPYLOBACTER INFECTIONS

Reducing the concentration and prevalence of *Campylobacter* in poultry and other animals at the farm and on the carcass is important to control campylobacteriosis. The use of specific phages and other biologic products to control the levels of *Campylobacter* (especially, antibiotic-resistant strains) colonizing chickens and other farm animals appear to be ideal prevention methods. There are still many challenges to achieving practical reduction of *Campylobacters* in poultry products.

The consumption of undercooked poultry and other meat products as well as contaminated water or milk continue to be major public health issues. More intensive surveillance, training, and education will be useful in reducing the staggering worldwide incidence of *Campylobacter* infections.

KEYWORDS

- *Campylobacter*
- foodborne pathogen
- gastroenteritis
- Guillain–Barré syndrome
- reactive arthritis
- adhesion
- invasion

REFERENCES

1. Sheppard, S. K.; Dallas, J. F.; Strachan, N. J. C.; MacRae, M.; McCarthy, N. D.; Wilson, D. J.; Gormley, F. J.; Falush, D.; Ogden, I. D.; Maiden, M. C.; Forbes, K. J. *Campylobacter* Genotyping to Determine the Source of Human Infection. *Clin. Infect. Dis.* **2009,** *48,* 1072–1078.
2. Scallan, E.; Hoekstra, R. M.; Angulo, F. J.; Tauxe, R. V.; Widdowson, M. A.; Roy, S. L.; Jones, J. L.; Griffin, P. M. Foodborne Illness Acquired in the United States—Major Pathogens. *Emerg. Infect. Dis.* **2011,** *17,* 7–15.
3. EFSA (European Food Safety Authority), ECDC (European Center for Disease Prevention and Control). The European Union Summary Report on Trends and Sources of Zoonoses, Zoonotic Agents and Food-borne Outbreaks in 2012. *EFSA J.* **2014,** *12,* 3547.
4. Oberhelman, R. A.; Taylor, D. N. *Campylobacter* Infection in Developing Countries. In *Campylobacter*, 2nd ed.; Nachamkin, I., Blaser, M. J., Eds.; ASM Press: Washington, DC, 2000; pp 139–153.
5. Nachamkin, I. *Campylobacter jejuni*. In *Food Microbiology: Fundamentals and Frontiers*, 3rd ed.; Doyle, M. P., Beuchat, L. R., Eds.; ASM Press: Washington, DC, 2007; Chapter 11, pp 237–248.
6. Silva, J.; Leite, D.; Fernandez, M.; Mena, C.; Gibbs, P. A.; Teixeira, P. *Campylobacter* spp. as a Foodborne Pathogen: A Review. *Ront. Microbiol.* **2011,** *2,* 1–12.
7. Skirrow, M. B. Diseases Due to Campylobacter, *Helicobacter* and Related Bacteria. *J. Comp. Pathol.* **1994,** *111,* 113–149.
8. Wilson, D. J.; Gabriel, E.; Leatherbarrow, A. J. H.; Cheesbrough, J.; Gee, S.; Bolton, E.; Fox, A.; Fearnhead, P.; Hart, C. A.; Diggle, P. J. Tracing the Source of Campylobacteriosis. *PLoS Genet.* **2008,** *4,* 1–9.
9. Altekruse, S. F.; Stern, N. J.; Fields, P. I.; Swerdlow, D. L. *Campylobacter jejuni*—An Emerging Foodborne Pathogen. *Emerg. Infect. Dis.* **1999,** *5,* 28–35.

10. Nachamkin, I.; Blaser, M. J.; Allos, B. M.; Ho, J. W. *Campylobacter jejuni* Infection and the Association with Guillain-Barre Syndrome. In *Campylobacter*, 2nd ed.; Nachamkin, I., Blaser, M. J., Eds.; ASM Press: Washington, DC, 2000; pp 155–175.
11. Ketley, J. M. Pathogenesis of Enteric Infection by *Campylobacter. Microbiology* **1997,** *143,* 5–21.
12. Jacobs-Reitsma, W. *Campylobacter* in the Food Supply. In *Campylobacter*, 2nd ed.; Nachamkin, I., Blaser, M. J., Eds.; ASM Press: Washington, DC, 2000; pp 467–481.
13. Reuter, M.; Mallett, A.; Pearson, B. M.; van Vliet, A. H. Biofilm Formation by *Campylobacter jejuni* is Increased Under Aerobic Conditions. *Appl. Environ. Microbiol.* **2010,** *76,* 2122–2128.
14. Nachamkin, I., Blaser, M. J., Tompkins, L. S., Eds. *Campylobacter jejuni* Current Status and Future Trends; ASM Press: Washington, DC, 1992.
15. Rollins, D. M.; Colwell, R. R. Viable but Nonculturable Stage of *Campylobacter jejuni* and Its Role in Survival in the Natural Aquatic Environment. *Appl. Environ. Microbiol.* **1986,** *52,* 531–538.
16. Levin, R. E. *Campylobacter jejuni*: A Review of its Characteristics, Pathogenicity, Ecology, Distribution, Subspecies Characterization and Molecular Methods of Detection. *Food Biotechnol.* **2007,** *21,* 271–347.
17. Humphrey, T. J. Injury and Recovery in Freeze- or Heat-damaged *Campylobacter jejuni. Lett. Appl. Microbiol.* **1986,** *3,* 81–84.
18. Hoffmann, S.; Batz, M. B.; Morris, J. G. Jr. Annual Cost of Illness and Quality-adjusted Life Year Losses in the United States due to 14 Foodborne Pathogens. *J. Food Prot.* **2012,** *75,* 1292–1302.
19. CDC Data on Foodborne Disease Outbreaks. Foodborne Disease Outbreak Surveillance System, 2009–2012. National Center for Emerging and Zoonotic Infectious Diseases, Division of Foodborne, Waterborne, and Environmental Diseases, 2014.
20. EFSA (European Food Safety Authority). Scientific Opinion on *Campylobacter* in Broiler Meat Production: Control Options and Performance Objectives and/or Targets at Different Stages of the Food Chain. *EFSA J.* **2011,** *9,* 2105.
21. Coker, A. O.; Isokpehi, R. D.; Thomas, B. N.; Amisu, K. O.; Obi, C. L. Human Campylobacteriosis in Developing Countries. *Emerg. Infect. Dis.* **2002,** *8,* 237–244.
22. Pitkänen, T. Review of *Campylobacter* spp. in Drinking and Environmental Waters. *J. Microbiol. Methods.* **2013,** *95,* 39–47.
23. Moore, J. E.; Corcoran, D.; Dooley, J. S.; Fanning, S.; Lucey, B.; Matsuda, M.; McDowell, D. A.; Mégraud, F.; Millar, B. C.; O'Mahony, R.; O'Riordan, L.; O'Rourke, M.; Rao, J. R.; Rooney, P. J.; Sails, A.; Whyte, P. *Campylobacter. Vet. Res.* **2005,** *36,* 351–382.
24. NARMS, *Retail Meat Complete Annual Report. Campylobacter* Data, FDA, 2011, pp 38–45.
25. Nauta, M. J.; van der Wal, F. J.; Putirulan, F. F.; Post, J.; van de Kassteele, J.; Bolder, N. M. "Evaluation of the Testing and Scheduling" Strategy for Control of *Campylobacter* in Broiler Meat in the Netherlands. *Int. J. Food Microbiol.* **2009,** *134,* 216–222.
26. Black, R. E.; Levine, M. M.; Clements, M. L.; Hughes, T. P.; Blaser, M. J. Experimental *Campylobacter jejuni* Infection in Humans. *J. Infect. Dis.* **1988,** *157,* 472–479.

27. Hu, L.; Kopecko, D. J. *Campylobacter* Species. In *International Handbook of Foodborne Pathogens;* Miliotis, M. D., Bier, J. W., Eds.; Marcel Dekker: New York, 2003; Chapter 12, pp 181–198.
28. Whiley, H.; van den Akker, B.; Giglio, S.; Bentham, R. The Role of Environmental Reservoirs in Human Campylobacteriosis. *Int. J. Environ. Res. Public. Health.* **2013,** *10,* 5886–5907.
29. Devane, M. L.; Moriarty, E. M.; Wood, D.; Webster-Brown, J.; Gilpin, B. J. The Impact of Major Earthquakes and Subsequent Sewage Discharges on the Microbial Quality of Water and Sediments in an Urban River. *Sci. Total Environ.* **2014,** *485–486,* 666–680.
30. Khan, I. U.; Gannon, V.; Jokinen, C. C.; Kent, R.; Koning, W.; Lapen, D. R.; Medeiros, D.; Miller, J.; Neumann, N. F.; Phillips, R.; Schreier, H.; Topp, E.; van Bochove, E.; Wilkes, G.; Edge, T. A. A National Investigation of the Prevalence and Diversity of Thermophilic *Campylobacter* Species in Agricultural Watersheds in Canada. *Water Res.* **2014,** *61C,* 243– 252.
31. Vogt, R. L.; Sours, H. E.; Barrett, T.; Feldman, R. A.; Dickinson, R. J.; Witherell, L. *Campylobacter enteritis* Associated with Contaminated Water. *Ann. Intern. Med.* **1982,** *96,* 292–296.
32. Wood, R. C.; MacDonald, K. L.; Osterholm, M. T. *Campylobacter* Enteritis Outbreaks Associated with Drinking Raw Milk during Youth Activities. A 10-Year Review of Outbreaks in the United States. *JAMA.* **1992,** *268,* 3228–3230.
33. FDA. Public Health Agencies Warn of Outbreaks Related to Drinking Raw Milk–Latest Outbreak of Campylobacteriosis in Midwest is Linked to Unpasteurized Product: US Food and Drug Administration, Silver Spring. [Online]. **2010**. http://www.fda.gov/NewsEvents/Newsroom/PressAnnouncements/ucm206311.htm. (accessed Sep 9, 2010).
34. Rapp, D.; Ross, C. M.; Pleydell, E. J.; Muirhead, R. W. Differences in the Fecal Concentrations and Genetic Diversities of *Campylobacter jejuni* Populations among Individual Cows in Two Dairy Herds. *Appl. Environ. Microbiol.* **2012,** *78,* 7564–7571.
35. Verhoeff-Bakkenes, L.; Jansen, H. A.; in't Veld, P. H.; Beumer, R. R.; Zwietering, M. H.; van Leusden, F. M. Consumption of Raw Vegetables and Fruits: A Risk Factor for *Campylobacter* Infections. *Int. J. Food Microbiol.* **2011,** *144,* 406–412.
36. Tauxe, R. V.; Hargrett-Bean, N.; Patton, C. M.; Wachsmuth, I. K. *Campylobacter* Isolates in the United States, 1982–1986. *MMWR CDC Surveill. Summ.* **1988,** *37,* 1–13.
37. Gillespie, I. A.; O'Brien, S. J.; Bolton, F. J. Age Patterns of Persons with Campylobacteriosis, England and Wales, 1990–2007. *Emerg. Infect. Dis.* **2009,** *15,* 2046–2048.
38. Taylor, D. N.; Pitarangsi, C.; Echeverria, P.; Diniega, B. M. *Campylobacter enteritis* during Doxycycline Prophylaxis for Malaria in Thailand [Letter]. *Lancet* **1988,** *2,* 578–579.
39. Skirrow, M. B. A Demographic Survey of *Campylobacter*, *Salmonella* and *Shigella* Infections in England. A Public Health Laboratory Service Survey. *Epidemiol. Infect.* **1987,** *9,* 647–657.
40. Friedman, C. R.; Neimann, J.; Wegener, H. C.; Tauxe, R. V. Epidemiology of *Campylobacter jejuni* Infections in the United States and Other Industrialized Nations. In *Campylobacter*, 2nd ed.; Nachamkin, I., Blaser, M. J., Eds.; ASM Press: Washington, DC, 2000; pp 121–138.

41. Unicomb, L. E.; Fullerton, K. E.; Kirk, M. D.; Stafford, R. J. Outbreaks of Campylobacteriosis in Australia, 2001 to 2006. *Foodborne Pathog. Dis.* **2009,** *6,* 1241–1250.
42. Nyachuba, D. G. Foodborne Illness: Is It on the Rise? *Nutr. Rev.* **2010,** *68,* 257–269.
43. Skirrow, M. B.; Jones, D. M.; Sutcliffe, E.; Benjamin, J. *Campylobacter* Bacteraemia in England and Wales, 1981–91. *Epidemiol. Infect.* **1993,** *110,* 567–573.
44. Kobayashi, R.; Matsumoto, S.; Yoshida, Y. Case of Acute Pancreatitis Associated with *Campylobacter* Enteritis. *World J. Gastroenterol.* **2014,** *20,* 7514–7517.
45. Blaser, M. J. *Campylobacter jejuni* and Related Species. In *Principles and Practice of Infectious Disease;* Mandell, G. L., Bennett, J. E., Dolin, R., Eds.; Churchill Livingstone: Philadelphia, PA, **2000**; pp 2276–2285.
46. Tanner, A. C.; Dzink, J. L.; Socransky, S. S.; Des Roches, C. L. Diagnosis of Periodontal Disease Using Rapid Identification of "Activity-related" Gram-negative Species. *J. Periodontal. Res.* **1987,** *22,* 207–220.
47. Keithlin, J.; Sargenant, J.; Thomas, M. K.; Fazil, A. Systematic Review and Meta-analysis of the Proportion of *Campylobacter* Cases that Develop Chronic Sequelae. *BMC Public Health.* **2014,** *14,* 1203.
48. Mughini Gras, L.; Smid, J. H.; Wagenaar, J. A.; de Boer, A. G.; Havelaar, A. H.; Friesema, I. H.; French, N. P.; Busani, L.; van Pelt, W. Risk Factors for Campylobacteriosis of Chicken, Ruminant, and Environmental Origin: A Combined Case-control and Source Attribution Analysis. *PLoS One* **2012,** *7,* e42599.
49. Peterson, M. C. Clinical Aspects of *Campylobacter jejuni* Infections in Adults. *West. J. Med.* **1994,** *161,* 148–152.
50. Allos, B. M. *Campylobacter jejuni* Infections: Update on Emerging Issues and Trends. *Clin. Infect. Dis.* **2001,** *32,* 1201–1216.
51. Altekruse, S. F.; Stern, N. J.; Fields, P. I.; Swerdlow, D. L. *Campylobacter jejuni* an Emerging Foodborne Pathogen. *Emerg. Infect. Dis.* **1999,** *5,* 28–35,
52. Nachamkin, I.; Allos, B. M.; Ho, J. W. *Campylobacter jejuni* Infection and the Association with Guillain-Barre Syndrome. In *Campylobacter*, 2nd ed.; Nachamkin, I., Blaser, M. J. Eds.; ASM Press: Washington, DC, **2000**; pp 155–175.
53. Jacobs, B. C.; Rothbarth, P. H.; van der Meche, F. G.; Herbrink, P.; Schmitz, P. I.; de Klerk, M. A.; van Doorn, P. A. The Spectrum of Antecedent Infections in Guillain-Barre' Syndrome: A Case-control Study. *Neurology.* **1998,** *51,* 1110–1115.
54. Moran, A. P. Structure and Conserved Characteristics of *Campylobacter jejuni* Lipopolysaccharides. *J. Infect. Dis.* **1997,** *176,* S115–S121.
55. Kuroki, S.; Saida, T.; Nukina, M.; Haruta, T.; Yoshioka, M.; Kobayashi, Y.; Nakanishi, H. *Campylobacter jejuni* Strains from Patients with Guillain-Barre' Syndrome Belong Mostly to Penner Serogroup 19 and Contain Beta-N-acetylglucosamine Residues. *Ann. Neurol.* **1993,** *33,* 243–247.
56. Lastovica, A. J.; Goddard, E. A.; Argent, A. C. Guillain-Barre' Syndrome in South Africa Associated with *Campylobacter jejuni* O:41 Strains. *J. Infect. Dis.* **1997,** *176,* S139–S143.
57. Mahendran, V.; Riordan, S. M.; Grimm, M. C.; Tran, T. A.; Major, J.; Kaakoush, N. O.; Mitchell, H.; Zhang, L. Prevalence of *Campylobacter* Species in Adult Crohn's Disease and the Preferential Colonization Sites of *Campylobacter* Species in the Human Intestine. *PLoS One* **2011,** *6,* e25417.

58. Mukhopadhya, I.; Thomson, J. M.; Hansen, R.; Berry, S. H.; El-Omar, E. M.; Hold, G. L. Detection of *Campylobacter concisus* and other *Campylobacter* Species in Colonic Biopsies from Adults with Ulcerative Colitis. *PLoS One* **2011,** *6* (6), e21490.
59. Griffiths, P. L.; Park, R. W. *Campylobacters* Associated with Human Diarrhoeal Disease. *J. Appl. Bacteriol.* **1990,** *69,* 281–301.
60. Parkhill, J.; Wren, B. W.; Mungall, K.; Ketley, J. M.; Churcher, C.; Basham, D.; Chillingworth, T.; Davies, R. M.; Feltwell, T.; Holroyd, S.; Jagels, K.; Karlyshev, A. V.; Moule, S.; Pallen, M. J.; Penn, C.W.; Quail, M. A.; Rajandream, M. A.; Rutherford, K. M.; van Vliet, A. H.; Whitehead, S.; Barrell, B. G. The Genome Sequence of the Foodborne Pathogen *Campylobacter jejuni* Reveals Hypervariable Sequences. *Nature* **2000,** *403,* 665–668.
61. Gundogdu, O.; Bentley, S. D.; Holden, M. T.; Parkhill, J.; Dorrell, N.; Wren, B. W. Reannotation and Re-analysis of the *Campylobacter jejuni* NCTC11168 Genome Sequence. *BMC Genomics* **2007,** *8,* 162.
62. Grant, C. C.; Konkel, M. E.; Cieplak, W. Jr.; Tompkins, L. S. Role of Flagella in Adherence, Internalization, and Translocation of *Campylobacter jejuni* in Nonpolarized and Polarized Epithelial Cell Cultures. *Infect. Immun.* **1993,** *61,* 1764–1771.
63. McSweegan, E.; Walker, R. I. Identification and Characterization of Two *Campylobacter jejuni* Adhesins for Cellular and Mucous Substrates. *Infect. Immun.* **1986,** *53,* 141–148.
64. Yao, R.; Burr, D. H.; Doig, P.; Trust, T. J.; Niu, H.; Guerry, P. Isolation of Motile and Nonmotile Insertional Mutants of *Campylobacter jejuni*: The Role of Motility in Adherence and Invasion of Eukaryotic Cells. *Mol. Microbiol.* **1994,** *14,* 883–893.
65. Pickett, C. L.; Pesci, E. C.; Cottle, D. L.; Russell, G.; Erdem, A. N.; Zeytin, H. Prevalence of Cytolethal Distending Toxin Production in *Campylobacter jejuni* and Relatedness of *Campylobacter* sp. *cdtB* Gene. *Infect. Immun.* **1996,** *64,* 2070–2078.
66. Martinez, I.; Meteo, E.; Churruca, E.; Girbau, C.; Alonso, R.; Fernández-Astorga, A. Detection of *cdtA*, *cdtB*, and *cdtC* Genes in *Campylobacter jejumi* by Multiplex PCR. *Int. J. Med. Microbiol.* **2006,** *296,* 45–48.
67. Abuoun, M.; Manning, G.; Cawthraw, S. A.; Ridley, A.; Ahmed, I. H.; Wassenaar, T. M.; Newell, D. G. Cytolethal Distending Toxin (CDT)-negative *Campylobacter jejuni* Strains and Anti-CDT Neutralizing Antibodies are Induced during Human Infection but not during Colonization in Chickens. *Infect. Immun.* **2005,** *73,* 3053–3062.
68. Jeon, B.; Itoh, K.; Ryu, S. Promoter Analysis of Cytolethal Distending Toxin Genes (*cdtA*, *B*, and *C*) and Effect of a *luxS* Mutation on CDT Production in *Campylobacter jejuni*. *Microbiol. Immunol.* **2005,** *49,* 599–603.
69. Smith, J. L.; Bayles, D. O. The Contribution of Cytolethal Distending Toxin to Bacterial Pathogenesis. *Crit. Rev. Microbiol.* **2006,** *32,* 227–248.
70. Hickey, T. E.; McVeigh, A. L.; Scott, D. A.; Michielutti, R. E.; Bixby, A.; Carroll, S. A.; Bourgeois, A. L.; Guerry, P. *Campylobacter jejuni* Cytolethal Distending Toxin Mediates Releases of Interleukin-8 from Intestinal Epithelial Cells. *Infect. Immun.* **2000,** *68,* 6535–6541.
71. Wassenaar, T. M. Toxin Production by *Campylobacter* spp. *Clin. Microbiol. Rev.* **1997,** *10,* 466–476.

72. Penner, J. L.; Hennessy, J. N. Passive Hemagglutination Technique for Serotyping *Campylobacter fetus* Subsp. *jejuni* on the Basis of Soluble Heat-stable Antigens. *J. Clin. Microbiol.* **1980,** *12,* 732–737.
73. Karlyshev, A. V.; Linton, D.; Gregson, N. A.; Lastovica, A. J.; Wren, B. W. Genetic and Biochemical Evidence of a *Campylobacter jejuni* Capsular Poly-saccharide that Accounts for Penner Serotype Specificity. *Mol. Microbiol.* **2000,** *35,* 529–541.
74. Guerry, P.; Poly, F.; Riddle, M.; Maue, A. C.; Chen, Y. H.; Monteiro, M. A. *Campylobacter* Polysaccharide Capsules: Virulence and Vaccines. *Front Cell Infect. Microbiol.* **2012,** *2,* 7.
75. Godschalk, P. C.; Heikema, A. P.; Gilbert, M.; Komagamine, T.; Ang, C. W.; Glerum, J.; Brochu, D.; Li, J.; Yuki, N.; Jacobs, B. C.; van Belkum, A.; Endtz, H. P. The Crucial Role of *Campylobacter jejuni* Genes in Anti-ganglioside Antibody Induction in Guillain-Barre Syndrome. *J. Clin. Invest.* **2004,** *114,* 1659–1665.
76. Allos, B. M.; Blaser, M. J. Potential Role of Lipopolysaccharides of *Campylobacter jejuni* in the Development of Guillain-Barre' Syndrome. *J. Endotoxin. Res.* **1995,** *2,* 237–238.
77. Goodyear, C. S.; O'Hanlon, G. M.; Plomp, J. J.; Wagner, E. R.; Morrison, I.; Veitch, J.; Cochrane L.; Bullens, R. W.; Molenaar, P. C.; Conner, J.; Willison, H. J. Monoclonal Antibodies Raised against GBS-associated *Campylobacter jejuni* Lipopolysaccharides React with Neuronal Gangliosides and Paralyze Muscle-nerve Preparations. *J. Clin. Invest.* **1999,** *104,* 697–708.
78. Koga, M.; Gilbert, M.; Li, J.; Koike, S.; Takahashi, M.; Furukawa, K.; Hirata, K.; Yuki, N. Antecedent Infections in Fisher Syndrome: A Common Pathogenesis of Molecular Mimicry. *Neurology* **2005,** *64,* 1605–1611.
79. Ang, C. W.; Jacobs, B. C.; Laman, J. D. The Guillain-Barre´ Syndrome: A True Case of Molecular Mimicry. *Trends Immunol.* **2004,** *25,* 61–66.
80. Ang, C. W.; Dijkstra, J. R.; de Klerk, M. A.; Endtz, H. P.; van Doorn, P. A.; Jacobs, B. C.; Jeurissen, S. H.; Wagenaar, J. A. Host Factors Determine Anti-GM1 Response Following Oral Challenge of Chickens with Guillain-Barré Syndrome Derived *Campylobacter jejuni* Strain GB11. *PLoS One* **2010,** *5,* e9820.
81. Bacon, D. J.; Szymanski, C. M.; Burr, D. H.; Silver, R. P.; Alm, R. A.; Guerry, P. A Phase Variable Capsule is Involved in Virulence of *Campylobacter jejuni* 81–176. *Mol. Microbiol.* **2001,** *40,* 769–777.
82. Rose, A.; Kay, E.; Wren, B. W.; Dallman, M. J. The *Campylobacter jejuni* NCTC 11168 Capsule Prevents Excessive Cytokine Production by Dendritic Cells. *Med. Microbiol. Immunol.* **2012,** *201,* 137–144.
83. Preston, M. A.; Penner, J. L. Characterization of Cross-reacting Serotypes of *Campylobacter jejuni. Can. J. Microbiol.* **1989,** *35,* 265–273.
84. Monterio, M. A.; Baqar, S.; Hall, E. R.; Chen, Y. H.; Porter, C. K.; Bentzel, D. E.; Applebee, L.; Guerry, P. Capsule Polysaccharide Conjugate Vaccine against Diarrheal Disease Caused by *Campylobacter jejuni. Infect. Immun.* **2009,** *77,* 1128–1136.
85. Maue, A. C.; Mohawk, K. L.; Giles, D. K.; Poly, F.; Ewing, C. P.; Jiao, Y.; Lee, G.; Ma, Z.; Monteiro, M. A.; Hill, C. L.; Ferderber, J. S.; Porter, C. K.; Trent, M. S.; Guerry, P. The Polysaccharide Capsule of *Campylobacter jejuni* Modulates the Host Immune Response. *Infect. Immue.* **2013,** *81,* 665–672.

86. Bachtiar, B. M.; Coloe, P. J.; Fry, B. N. Knockout Mutagenesis of the *kpsE* Gene of *Campylobacter jejuni* 81116 and Its Involvement in Bacterium-host Interactions. *FEMS Immunol. Med. Microbiol.* **2007,** *49,* 149–154.
87. Fry, B. N.; Feng, S.; Chen, Y. Y.; Newell, D. G.; Coloe, P. J.; Korolik, V. The *galE* Gene of *Campylobacter jejuni* is Involved in Lipopolysaccharide Synthesis and Virulence. *Infect. Immun.* **2000,** *68,* 2594–2601.
88. Szymanski, C. M.; Burr, D. H.; Guerry, P. *Campylobacter* Protein Glycosylation Affects Host Cell Interactions. *Infect. Immun.* **2002,** *70,* 2242–2244.
89. Stahl, M.; Ries, J.; Vermeulen, J.; Yang, H.; Sham, H. P.; Crowley, S. M.; Badayeva, Y.; Turvey, S. E.; Gaynor, E. C.; Li, X.; Vallance, B. A. A Novel Mouse Model of *Campylobacter jejuni* Gastroenteritis Reveals Key Pro-inflammatory and Tissue Protective Roles for Toll-like Receptor Signaling during Infection. *PLoS Pathog.* **2014,** *10,* e1004264.
90. Bacon, D. J.; Alm, R. A.; Hu, L.; Hickey, T. E.; Ewing, C. P.; Batchelor, R. A.; Trust, T. J.; Guerry, P. DNA Sequence and Mutational Analyses of the *pVir* Plasmid of *Campylobacter jejuni* 81–176. *Infect. Immun.* **2002,** *70,* 6242–6250.
91. Bleumink-Pluym, N. M.; van Alphen, L. B.; Bouwman, L. I.; Wösten, M. M.; van Putten, J. P. Identification of a Functional Type VI Secretion System in *Campylobacter jejuni* Conferring Capsule Polysaccharide Sensitive Cytotoxicity. *PLoS Pathog.* **2013,** *9,* e1003393.
92. Buelow, D. R.; Christensen, J. E.; Neal-McKinney, J. M.; Konkel, M. E. *Campylobacter jejuni* Survival within Human Epithelial Cells is Enhanced by the Secreted Protein CiaI. *Mol. Microbiol.* **2011,** *80,* 1296–1312.
93. Neal-McKinney, J. M.; Konkel, M. E. The *Campylobacter jejuni* CiaC Virulence Protein is Secreted from the Flagellum and Delivered to the Cytosol of Host Cells. *Front. Cell. Infect. Microbiol.* **2012,** *2,* 31.
94. Hu, L.; Kopecko, D. J. Cell Biology of Human Host Cell Entry by *Campylobacter jejuni*. In *Campylobacter*, 3rd ed.; Nachamkin, I., Szymanski, C. M., Blaser, M. J., Eds.; ASM Press: Washington, DC, 2008; pp 297–313.
95. Novik, V.; Hofreuter, D.; Galán, J. E. Identification of *Campylobacter jejuni* Genes Involved in Its Interaction with Epithelial Cells. *Infect. Immun.* **2010,** *78,* 3540–3553.
96. Flanagan, R. C.; Neal-McKinney, J. M.; Dhillon, A. S.; Miller, W. G.; Konkel, M. E. Examination of *Campylobacter jejuni* Putative Adhesins Leads to the Identification of a New Protein, Designated FlpA, Required for Chicken Colonization. *Infect. Immun.* **2009,** *77,* 2399–2407.
97. Ashgar, S. S.; Oldfield, N. J.; Wooldridge, K. G.; Jones, M. A.; Irving, G. J.; Turner, D. P.; Ala'Aldeen D. A. CapA, an Autotransporter Protein of *Campylobacter jejuni*, Mediates Association with Human Epithelial Cells and Colonization of the Chicken Gut. *J. Bacteriol.* **2007,** *189,* 1856–1865.
98. Javed, M. A.; Grant, A. J.; Bagnall, M. C.; Maskell, D. J.; Newell, D. G.; Manning, G. Transposon Mutagenesis in a Hyper-invasive Clinical Isolate of *Campylobacter jejuni* Reveals a Number of Genes with Potential Roles in Invasion. *Microbiology* **2010,** *156,* 1134–1143.
99. Jin, S.; Joe, A.; Lynett, J.; Hani, E. K.; Sherman, P.; Chan, V. L. JlpA, a Novel Surface-exposed Lipoprotein Specific to *Campylobacter jejuni*, Mediates Adherence to Host Epithelial Cells. *Mol. Microbiol.* **2001,** *39,* 1225–1236.

100. Kervella, M.; Pages, J. M.; Pei, Z.; Grollier, G.; Blaser, M. J.; Fauchere, J. L. Isolation and Characterization of Two *Campylobacter* Glycine-extracted Proteins that Bind to HeLa Cell Membranes. *Infect. Immun.* **1993,** *61,* 3440–3448.
101. Konkel, M. E.; Kim, B. J.; Rivera-Amill, V.; Garvis, S. G. Bacterial Secreted Proteins are Required for the Internalization of *Campylobacter jejuni* into Cultured Mammalian Cells. *Mol. Microbiol.* **1999,** *32,* 691–701.
102. Pei, Z. H.; Ellison, R. T. 3rd.; Blaser, M. J. Identification, Purification, and Characterization of Major Antigenic Proteins of *Campylobacter jejuni*. *J. Biol. Chem.* **1991,** *266,* 16363–16369.
103. Hofreuter, D.; Novik, V.; Galán, J. E. Metabolic Diversity in *Campylobacter jejuni* Enhances Specific Tissue Colonization. *Cell Host. Microbe.* **2008,** *45,* 425–433.
104. Leon-Kempis Mdel, R.; Guccione, E.; Mulholland, F.; Williamson, M. P.; Kelly, D. J. The *Campylobacter jejuni* PEB1a Adhesin is an Aspartate/Glutamatebinding Protein of an ABC Transporter Essential for Microaerobic Growth on Dicarboxylic Amino Acids. *Mol. Microbiol.* **2006,** *60,* 1262–1275.
105. Hofreuter, D.; Mohr, J.; Wensel, O.; Rademacher, S.; Schreiber, K.; Schomburg, D.; Gao, B.; Galán, J. E. Contribution of Amino Acid Catabolism to the Tissue Specific Persistence of *Campylobacter jejuni* in a Murine Colonization Model. *PLoS One* **2012,** *7,* e50699.
106. Joshua, G. W.; Guthrie-Irons, C.; Karlyshev, A. V.; Wren, B. W. Biofilm Formation in *Campylobacter jejuni*. *Microbiology* **2006,** *152,* 387–396.
107. Bronnec, V.; Turoňová, H.; Bouju, A.; Cruveiller, S.; Rodrigues, R.; Demnerova, K.; Tresse, O.; Haddad, N.; Zagorec, M. Adhesion, Biofilm Formation, and Genomic Features of *Campylobacter jejuni* Bf, an Atypical Strain Able to Grow under Aerobic Conditions. *Front Microbiol.* **2016,** *7,* 1002.
108. Kalmokoff, M.; Lanthier, P.; Tremblay, T. L.; Foss, M.; Lau, P. C.; Sanders, G.; Austin, J.; Kelly, J.; Szymanski, C. M. Proteomic Analysis of *Campylobacter jejuni* 11168 Biofilms Reveals a Role for the Motility Complex in Biofilm Formation. *J. Bacteriol.* **2006,** *188,* 4312–4320.
109. Reeser, R. J.; Medler, R. T.; Billington, S. J.; Jost, B. H.; Joens, L. A. Characterization of *Campylobacter jejuni* Biofilms under Defined Growth Conditions. *Appl. Environ. Microbiol.* **2007,** *73,* 1908–1913.
110. Rubinchik, S.; Seddon, A. M.; Karlyshev, A. V. A Negative Effect of *Campylobacter* Capsule on Bacterial Interaction with an Analogue of a Host Cell Receptor. *BMC Microbiol.* **2014,** *14,* 141.
111. Kopecko, D. J.; Hu, L.; Zaal, K. J. *Campylobacter jejuni*-microtubule Dependent Invasion. *Trends Microbiol.* **2001,** *9,* 389–396.
112. Hu, L.; Raybourne, R. B.; Kopecko, D. J. Ca^{2+} Release from Host Intracellular Store and Related Signal Transduction during *Campylobacter jejuni* 81–176 Internalization into Human Intestinal Cells. *Microbiology* **2005,** *151,* 3097–3105.
113. Hu, L.; McDaniel, J.; Kopecko, D. J. Signal Transduction Events Involved in Human Epithelial Cell Invasion by *Campylobacter jejuni* 81–176. *Microb. Pathog.* **2006a,** *40,* 91–100.
114. Krause-Gruszczynska, M.; Rohde, M.; Hartig, R.; Genth, H.; Schmidt, G.; Keo, T.; König, W.; Miller, W. G.; Konkel, M. E.; Backert, S. Role of the Small Rho GTPases

Rac1 and Cdc42 in Host Cell Invasion of *Campylobacter jejuni*. *Cell Microbiol.* **2007,** *9,* 2431–2444.

115. Krause-Gruszczynska, M.; Boehm, M.; Rohde, M.; Tegtmeyer, N.; Takahashi, S.; Buday, L.; Oyarzabal, O. A.; Backert, S. The Signaling Pathway of *Campylobacter jejuni*-induced Cdc42 Activation: Role of Fibronectin, Integrin Betal, Tyrosine Kinases and Guanine Exchange Factor Vav2. *Cell Commun. Signal.* **2011,** *9,* 32.
116. Boehm, M.; Hoy, B.; Rohde, M.; Tegtmeyer, N.; Baek, K. T.; Oyarzabal, O. A.; Brondsted, L.; Wessler, S.; Backert, S. Rapid Paracellular Transmigration of *Campylobacter jejuni* Across Polarized Epithelial Cells without Affecting TER: Role of Proteolytic-active HtrA Cleaving E-cadherin but not Fibronectin. *Gut Pathog.* **2012,** *4,* 3.
117. Hu, L.; Tall, B. D.; Curtis, S. K.; Kopecko, D. J. Enhanced Microscopic Definition of *Campylobacter jejuni* 81–176 Adherence to, Invasion into, Translocation Across, and Exocytosis from Polarized Human Intestinal Caco-2 Cells. *Infect. Immun.* **2008,** *76,* 5294–5304.
118. Alemka, A.; Clyne, M.; Shanahan, F.; Tompkins, T.; Corcionivoschi, N.; Bourke, B. Probiotic Colonization of the Adherent Mucus Layer of HT29MTXE12 Cells Attenuates *Campylobacter jejuni* Virulence Properties. *Infect. Immun.* **2010,** *78,* 2812–2822.
119. Brondsted, L.; Andersen, M. T.; Parker, M.; Jørgensen, K.; Ingmer, H. The HtrA Protease of *Campylobacter jejuni* is Required for Heat and Oxygen Tolerance and for Optimal Interaction with Human Epithelial Cells. *Appl. Environ. Microbiol.* **2007,** *71,* 3205–3212.
120. Bæk, K. T.; Vegge, C. S.; Brøndsted, L. HtrA Chaperone Activity Contributes to Host Cell Binding in *Campylobacter jejuni*. *Gut Pathog.* **2011,** *3,* 13.
121. Hoy, B.; Geppert, T.; Boehm, M.; Reisen, F.; Plattner, P.; Gadermaier, G.; Sewald, N.; Ferreira, F.; Briza, P.; Schneider, G.; Backert, S.; Wessler, S. Distinct roles of Secreted HtrA Proteases from Gram-negative Pathogens in Cleaving the Junctional Protein and Tumor Suppressor E-cadherin. *J. Biol. Chem.* **2012,** *287,* 10115–10120.
122. Kiehlbauch, J. A.; Albach, R. A.; Baum, L. L.; Chang, K. P. Phagocytosis of *Campylobacter jejuni* and Its Intracellular Survival in Mononuclear Phagocytes. *Infect. Immun.* **1985,** *48,* 446–451.
123. Hickey, T. E.; Majam, G.; Guerry, P. Intracellular Survival of *Campylobacter jejuni* in Human Monocytic Cells and Induction of Apoptotic Death by Cytholethal Distending Toxin. *Infect. Immun.* **2005,** *73,* 5194–5197.
124. Day, W. A. Jr.; Sajecki, J. L.; Pitts, T. M.; Joens, L. A. Role of Catalase in *Campylobacter jejuni* Intracellular Survival. *Infect. Immun.* **2000,** *68,* 6337–6345.
125. Watson R. O.; Galan, J. E. *Campylobacter jejuni* Survives within Epithelial Cells by Avoiding Delivery to Lysosomes. *PLoS Pathog.* **2008,** *4,* e14.
126. Hickey, T. E.; Baqar, S.; Bourgeois, A. L.; Ewing, C. P.; Guerry, P. *Campylobacter jejuni* Stimulated Secretion of Interleukin-8 by INT407 Cells. *Infect. Immun.* **1999,** *67,* 88–93.
127. Hu, L.; Bray, M. D.; Osorio, M.; Kopecko, D. J. *Campylobacter jejuni* Induced Maturation and Cytokine Production in Human Dendritic Cells. *Infect. Immun.* **2006b,** *74,* 2697–7205.

128. Hu, L.; Bray, M. D.; Geng, Y.; Kopecko, D. J. *Campylobacter jejuni*-mediated Induction of CC and CXC Chemokines and Chemokine Receptors in Human Dendritic Cells. *Infect. Immun.* **2012,** *80,* 2929–2939.
129. Crocker, P. R.; Paulson, J. C.; Varki, A. Siglecs and Their Roles in the Immune System. *Nat. Rev. Immunol.* **2007,** *7,* 255–266.
130. Stephenson, H. N.; Mills, D. C.; Jones, H.; Milioris, E.; Copland, A.; Dorrell, N.; Wren, B. W.; Crocker, P. R.; Escors, D.; Bajaj-Elliott, M. Pseudaminic Acid on *Campylobacter jejuni* Flagella Modulates Dendritic Cell IL-10 Expression via Siglec-10 Receptor: A Novel Flagellin-Host Interaction. *J. Infect. Dis.* **2014,** *210,* 1487–1498.
131. Sheppard, S. K.; Jolley, K. A.; Maiden, M. C. A Gene-by-gene Approach to Bacterial Population Genomics: Whole Genome MLST of *Campylobacter*. *Genes (Basel).* **2012,** *3,* 261–277.
132. Taboada, E. N.; Clark, C. G.; Sproston, E. L.; Carrillo, C. D. Current Methods for Molecular Typing of *Campylobacter* Species. *J. Microbiol. Methods.* **2013,** *95,* 24–31.
133. On, S. L. Isolation, Identification and Subtyping of *Campylobacter*: Where to from Here? *J. Microbiol. Methods.* **2013,** *95,* 3–7.
134. Kulkarni, S.; Lever, S.; Logan, J. M. J. Detection of *Campylobacter* Species: A Comparison of Culture and Polymerase Chain Reaction Based Methods. *J. Clin. Pathol.* **2002,** *55,* 749–753.
135. ISO 17995. Water Quality. Detection and Enumeration of Thermotolerant *Campylobacter* Species. International Organization for Standardization, Geneva, Switzerland, 2005.
136. Moore, J.; Caldwell, P.; Millar, B. Molecular Detection of *Campylobacter* spp. in Drinking, Recreational and Environmental Water Supplies. *Int. J. Hyg. Environ. Health.* **2001,** *204,* 185–189.
137. Poly, F.; Serichatalergs, O.; Schulman, M.; Ju, J.; Cates, C. N.; Kanipes, M. I.; Mason, C.; Guerry, P. Discrimination of Major Capsular Types of *Campylobacter jejuni* by Multiplex PCR. *J. Clin. Microbiol.* **2011,** *49,* 1750–1757.
138. Alexandrino, M.; Grohmann, E.; Szewzyk, U. Optimization of PCR-based Methods for Rapid Detection of *Campylobacter jejuni, Campylobacter coli* and *Yersinia enterocolitica* Serovar 0:3 in Wastewater Samples. *Water Res.* **2004,** *38,* 1340–1346.
139. LaGier, M. J.; Joseph, L. A.; Passaretti, T. V.; Musser, K. A.; Cirino, N. M. A Real-time Multiplexed PCR Assay for Rapid Detection and Differentiation of *Campylobacter jejuni* and *Campylobacter coli. Mol. Cell Probes.* **2004,** *18,* 275–282.
140. Seinige, D.; von Köckritz-Blickwede, M.; Krischek, C.; Klein, G.; Kehrenberg, C. Influencing Factors and Applicability of the Viability EMA-qPCR for a Detection and Quantification of *Campylobacter* Cells from Water Samples. *PLoS One* **2014,** *9,* e113812.
141. Steensels, D.; Verhaegen, J.; Lagrou, K. Matrix-assisted Laser Desorption Ionization-time of Flight Mass Spectrometry for the Identification of Bacteria and Yeasts in a Clinical Microbiological Laboratory: A Review. *Acta Clin. Belg.* **2011,** *66,* 267–273.
142. Oyarzabal, O. A.; Battie, C. Immunological Methods for the Detection of *Campylobacter* spp. Current Applications and Potential Use in Biosensors. In *Trends in Immunolabelled and Related Techniques*; Abuelzein, E., Ed.; InTech: London, UK, 2012.

143. Yang, X.; Kirsch, J.; Simonian, A. *Campylobacter* spp. Detection in the 21st Century: A Review of the Recent Achievements in Biosensor Development. *J. Microbiol. Methods.* **2013,** *95,* 48–56.
144. Tauxe, R. V. Emerging Foodborne Pathogens. *Int. J. Food Microbiol.* **2002,** *78,* 31–41.
145. Monk, A. B.; Rees, C. D.; Barrow, P.; Hagens, S.; Harper, D. R. Bacteriophage Applications: Where are we now? *Lett. Appl. Microbiol.* **2010,** *51,* 363–369.
146. Svensson, S. L.; Davis, L. M.; MacKichan, J. K.; Allan, B. J.; Pajaniappan, M.; Thompson, S. A.; Gaynor E. C. The CprS Sensor Kinase of the Zoonotic Pathogen *Campylobacter jejuni* Influences Biofilm Formation and is Required for Optimal Chick Colonization. *Mol. Microbiol.* **2009,** *71,* 253–272.
147. Fields, J. A.; Thompson, S. A. *Campylobacter jejuni* CsrA Mediates Oxidative Stress Responses, Biofilm Formation, and Host Cell Invasion. *J. Bacteriol.* **2008,** *190,* 3411–3416.
148. Reuter, M.; Mallett, A.; Pearson, B. M.; van Vliet, A. H. Biofilm Formation by *Campylobacter jejuni* is Increased under Aerobic Conditions. *Appl. Environ. Microbiol.* **2010,** *76,* 2122–2128.
149. Naito, M.; Frirdich, E.; Fields, J. A.; Pryjm,a M.; Li, J.; Cameron, A.; Gilbert, M.; Thompson, S. A.; Gaynor E. C. Effects of Sequential *Campylobacter jejuni* 81–176 lipooligosaccharide Core Truncations on Biofilm Formation, Stress Survival, and Pathogenesis. *J. Bacteriol.* **2010,** *192,* 2182–2192.
150. Drozd, M.; Gangaiah, D.; Liu, Z.; Rajashekara, G. Contribution of TAT System Translocated PhoX to *Campylobacter jejuni* Phosphate Metabolism and Resilience to Environmental Stresses. *PLoS One.* **2011,** *6,* e26336.
151. Gangaiah, D.; Liu, Z.; Arcos, J.; Kassem, I. I.; Sanad, Y.; Torrelles, J. B.; Rajashekara, G. Polyphosphate Kinase 2: A Novel Determinant of Stress Responses and Pathogenesis in *Campylobacter jejuni. PLoS One* **2010,** *5,* e12142.
152. Guerry, P.; Ewing, C. P.; Schirm, M.; Lorenzo, M.; Kelly, J.; Pattarini, D.; Majam, G.; Thibault, P.; Logan, S. Changes in Flagellin Glycosylation Affect *Campylobacter* Autoagglutination and Virulence. *Mol. Microbiol.* **2006,** *60,* 299–311.
153. Rajashekara, G.; Drozd, M.; Gangaiah, D.; Jeon, B.; Liu, Z.; Zhang, Q. Functional Characterization of the Twin-arginine Translocation System in *Campylobacter jejuni. Foodborne Pathog. Dis.* **2009,** *6,* 935–945.
154. Sampathkumar, B.; Napper, S.; Carrillo, C. D.; Willson, P.; Taboada, E.; Nash, J. H.; Potter, A. A.; Babiuk, L. A.; Allan, B. J. Transcriptional and Translational Expression Patterns Associated with Immobilized Growth of *Campylobacter jejuni. Microbiology* **2006,** *152,* 567–577.

CHAPTER 4

ESCHERICHIA COLI O157:H7

XIANGNING BAI and YANWEN XIONG*

Collaborative Innovation Center for Diagnosis and Treatment of Infectious Diseases, State Key Laboratory of Infectious Disease Prevention and Control, National Institute for Communicable Disease Control and Prevention, Chinese Center for Disease Control and Prevention, Beijing, PR China

**Corresponding author. E-mail: xiongyanwen@icdc.cn*

CONTENTS

Abstract 94
4.1 Introduction 94
4.2 Biological Characteristics 95
4.3 Diseases 96
4.4 Epidemiology 97
4.5 Genomics 100
4.6 Pathogenesis 102
4.7 Identification and Detection 107
4.8 Treatment and Prevention 112
4.9 Future Perspective 115
Keywords 116
References 116

ABSTRACT

Escherichia coli O157:H7 was first identified as a pathogen associated with hemorrhagic colitis (HC) in 1982 in an outbreak of foodborne-related illness after consuming undercooked ground hamburger meat in a fast-food chain in the United States. Since then, *E. coli* O157:H7 has been reported to be responsible for numerous infectious outbreaks worldwide. Infections are associated with the ingestion of contaminated food and water, as well as person-to-person contact, and through contact with animals or their environment. The produce of Shiga toxin (Stx) is the major virulence feature attributed to *E. coli* O157:H7 disease pathogenesis. This chapter provides a board overview of *E. coli* O157:H7 with an emphasis on the epidemiology, pathogenesis, detection, treatment, and prevention.

4.1 INTRODUCTION

Escherichia coli (*E. coli*) are Gram-negative, facultative anaerobic and rod-shaped bacteria of the genus *Escherichia* that normally live in the intestines of humans and warm-blooded animals. Most *E. coli* strains are harmless and actually are important part of normal flora in the intestinal tract of human and animals. However, some *E. coli* strains are pathogenic through gene gain and loss, which could cause a broad range of diseases that span from the gastrointestinal tract to extraintestinal sites such as urinary tract, respiratory tract, bloodstream, and central nervous system. Pathogenic *E. coli* strains are further categorized into different pathotypes or pathovars based on distinct pathogenic features. Six major pathotypes are associated with diarrhea and collectively are referred to as diarrheagenic *E. coli* (DEC), including Shiga toxin-producing *E. coli* (STEC) or enterohemorrhagic *E. coli* (EHEC), enterotoxigenic *E. coli* (ETEC), enteropathogenic *E. coli* (EPEC), enteroaggregative *E. coli* (EAEC), enteroinvasive *E. coli* (EIEC), and diffusely adherent *E. coli* (DAEC).[1]

Some kinds of *E. coli* cause disease by producing a toxin called Shiga toxin (Stx). Stx can kill cultured Vero cells and was also called Verocytotoxin. The bacteria that produce these toxins are called STEC or Verocytotoxin-producing *E. coli* (VTEC). EHEC is a subset of STEC and was originally described by its association with HC. *E. coli* O157 was first identified as a pathogen in 1982 in USA. It was isolated from 47 persons who had developed bloody diarrhea after consuming hamburgers contaminated with this organism.[2] Since then, *E. coli* O157 has been identified in many other sources, and most of them were from outbreak investigations.[3] *E. coli* O157 can be transmitted to humans through contaminated food and water, through contact with animals or their environment, and directly through person to person. The most common reservoir is cattle, and ground beef is the most frequently identified vehicle of transmission to humans. Despite over 400 STEC serotypes have been identified already, the most common STEC serotype is O157:H7. *E. coli* O157 strains have been considered to be most virulent and extensively studied worldwide. In addition to *E. coli* O157, six EHEC serogroups, that is, O26, O45, O103, O111, O121, and O145 (known as the "Big 6") are the most commonly non-O157 STEC serogroups associated with severe illness in humans reported in many countries. However, less is known about the non-O157 STECs, especially, those less predominant serotypes, because of their high genotypic and phenotypic diversity. To date, there is no standard detection method covering all STEC serotypes and identification of non-O157 STEC infections is much more complex. Hence, the non-O157 serogroups are regarded less likely to cause severe illness than *E. coli* O157, though sometimes they can.

4.2 BIOLOGICAL CHARACTERISTICS

E. coli is the type species in the genus *Escherichia*, which mostly contains motile Gram-negative bacilli within the family Enterobacteriaceae and the tribe *Escherichia. E. coli* can be recovered easily from clinical specimens on general or selective media at 37°C under aerobic conditions. *E. coli* O157:H7 can grow in temperatures ranging from 7 to 43°C, with an optimum temperature of 37°C. Some *E. coli* O157:H7 strains can grow in acidic foods, down to a pH of 4.4, and in foods with a minimum water activity (Aw) of 0.95. It is destroyed by thorough cooking of foods until all parts reach a temperature of 70°C or higher. Selective media like Sorbitol MacConkey agar for isolating *E. coli* O157:H7 strains rely on the fact that most of these strains display characteristic biochemical reactions: no β-glucuronidase activity and no D-sorbitol fermentation within 24 h at 37°C.

4.3 DISEASES

While individuals may be asymptomatic even though *E. coli* O157:H7 is detected in stools, others can develop severe cases of infection characterized by abdominal cramps and bloody diarrhea within three days of consuming contaminated foodstuffs.[4] People of any age can get infected, but the most severe cases of infection typically occur in children <5 years of age, the elderly >65 years old and in immune-compromised persons. The symptoms of *E. coli* O157:H7 infections vary for each person but often include severe stomach cramps, diarrhea (often bloody), and vomiting. If there is fever, it is usually not very high (less than 101°F/38.5°C). Most people get better within 5–7 days. Some infections are very mild, but others are severe or even life-threatening. Around 5–10% of those who are diagnosed with *E. coli* O157:H7 infections tend to develop a potentially life-threatening complication known as hemolytic-uraemic syndrome (HUS). HUS is characterized by acute renal failure, hemolytic anemia, and thrombocytopenia. Clues that a person is developing HUS include decreased frequency of urination, feeling very tired, and losing pink color in cheeks and inside the lower eyelids. Persons with HUS should be hospitalized because their kidneys may stop working and they may develop serious syndrome. Most people with HUS recover within a few weeks, but some may suffer permanent damage or even death. Overall, HUS is the most common cause of acute renal failure in young children. It can cause neurological complications (such as seizure, stroke, and coma) in 25% of HUS patients and chronic renal sequelae, usually mild, in around 50% of survivors.[5,6]

The time between ingesting the *E. coli* O157:H7 bacteria and feeling sick is usually 3–4 days after the exposure, but may be as short as 1 day or as long as 10 days. The symptoms often begin slowly with mild belly pain or non-bloody diarrhea and then worsen over several days. HUS, if it occurs, develops an average 7 days after the first symptoms.

The very young and the elderly are most susceptible to complications and death from *E. coli* O157:H7 infection. This bacterial infection is the leading cause of acute kidney failure in healthy children in developed countries. Children younger than 5 years old are more likely to develop complications requiring hospitalization and kidney dialysis, and those older than 60 years are more likely to die regardless of the clinical complications.[7] Long-term complications after HUS are encountered frequently, including chronic renal insufficiency, neurologic disorders, hypertension, and other cardiovascular diseases. Late gastrointestinal sequelae are common in patients with *E. coli* O157:H7 infection even without HUS.[8]

4.4 EPIDEMIOLOGY

Despite EHEC O157:H7 is a human pathogen responsible for numerous infectious outbreaks worldwide, it is also a resident commensal bacterium commonly found in the intestinal tract of ruminants such as cattle, sheep, goats, and deer. Human exposure to this microbial pathogen is classically associated with the ingestion of undercooked ground beef;[9] however, infections can also arise following the ingestion of ruminant feces-contaminated foodstuffs such as fruits, vegetables, and drinking water, as well as person-to-person contact and hospital acquired nosocomial infections[10] (Fig. 4.1).

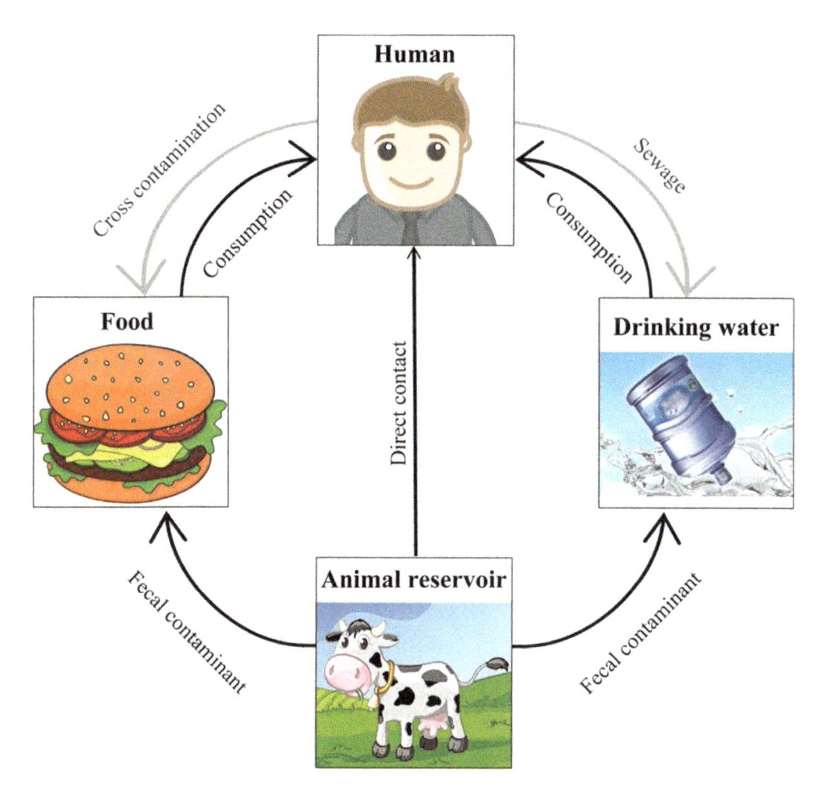

FIGURE 4.1 Transmission modes for *E. coli* O157:H7.

The natural reservoir of *E. coli* O157:H7 is the gastrointestinal tract of cattle and other ruminant animals. Animal feces can contaminate food, irrigation water, or drinking water. Humans become exposed following the ingestion of contaminated food or water or through direct contact with colonized animals.

E. coli O157:H7 outbreaks are attributed to its low infectious dose (<100 organisms) and high transmissibility, which can either remain isolated or develop into widespread international outbreaks. In 2000, one of the largest EHEC outbreak in Canada occurred in Walkerton, Ontario, due to inadequately chlorinated drinking water, which resulted in approximately 2300 cases of infection and seven deaths.[11] Another human outbreak occurred in Canada in 2012 due to EHEC O157:H7 infection arising from the fecal contamination of huge volumes of meat in a single processing plant situated in Southern Alberta. In North America, approximately 75,000 cases of EHEC infections are reported annually. Of these, approximately 10–15% developed HUS, an additional 5–10% resulted in long-term complications and 3–5% of HUS cases were fatal.[12] In China, *E. coli* O157:H7 were detected in different animals and caused a major outbreak in 1999 causing 195 HUS cases and 177 deaths.[13,14]

The natural reservoir of STEC is the gastrointestinal tract of cattle and other animals. The pathogen does not cause disease in these animals that seem to lack receptors for Stxs.[15] The most common means of spread, particularly in outbreak situations, is ingestion of contaminated food products. Consumption of undercooked ground beef has been implicated in many major outbreaks and was also the culprit in the first described outbreak of *E. coli* O157:H7 in 1982, affecting 47 patients in the Pacific Northwest.[2] Other types of meat, unpasteurized dairy products, and leafy green vegetables are also commonly implicated as sources of outbreaks.[16–18] Almost any food product can be contaminated, as exemplified by outbreaks caused by apple juice, sprouts, and more recently, 77 patients in the United States who were infected through consumption of ready-to-bake cookie dough.[19]Foods typically are contaminated through the slaughter and processing of colonized animals, shedding of pathogens from colonized cattle into milk, and use of contaminated soil or contaminated irrigation water in produce production. Table 4.1 summarizes some outbreaks of *E. coli* O157:H7 in the United States and worldwide in recent years.

TABLE 4.1 Selected Outbreaks of *E. coli* O157:H7 in the United States and Worldwide.

Year	Country	Cases	HUS	Source	References
2007	United States	7	0	Petting zoo	[20]
2007	Scotland	9	0	Cold meat salad	[20]
2007	Netherlands	41	0	Lettuce	[20]
2007	Iceland	9	0	Lettuce	[20]
2008	Canada	235	1	Raw onions	[20]
2008	United States	49	1	Ground beef	[20]
2009	United States	72	10	Dough	[20]
2009	United Kingdom	4	2	Fast food outlet	[20]
2009	Germany	12	4	Playground	[21]
2011	Japan	181	34	Raw beef dishes	[22]
2011	United States	58	4	Romaine lettuce	[23]
2011	Japan	142	25	Rice cakes	[24]
2012	United States	17	0	Bagged salad	[25]
2012	United Kingdom	38	0	Person-to-person	[26]
2012	Japan	54	5	Pickles	[27]
2013	United States	33	2	Salads	[US CDC]
2013	Canada	7	1	Beef and veal tartares	[28]
2014	United States	12	0	Ground beef	[US CDC]
2014	United Kingdom	15	5	Lamb-feeding	[29]
2015	United States	19	0	Chicken salad	(US CDC)
2016	United States	11	0	Sprouts	(US CDC)

Disease caused by *E. coli* O157:H7 has a global impact and has been reported from all six populated continents. Data on population-level incidence of sporadic STEC infections worldwide are scarce, but estimates range between 0.6 and 136 cases per 100,000 patient-years in different countries, with up to one-third of cases caused by STEC O157. A previous systematic review by Majowicz et al., estimated that STEC infection causes 2.8 million acute illnesses a year on a global scale. [30] Multiple large outbreaks have been recorded since 1982, mostly in industrialized countries. This might be a result of reporting bias owing to diagnostic resources and national reporting systems available in the developed world; however, other factors are contributing to this differential occurrence in outbreaks, including the large-scale centralized food production in industrialized countries that can serve as a widespread vector of dissemination in cases of food contamination.

In addition to serotype O157:H7, multiple non-O157 STEC also can cause human infections and disease outbreaks. In recent years, there is growing concern over the emergence of more than 200 non-O157 STEC serotypes associated with human illness, including the life-threatening HUS.[31] In fact, non-O157 STEC strains are responsible for a larger portion of STEC infections than O157 strains in the United States, Canada, Australia, Latin America, and Europe.[32–34] Our previous studies also showed that healthy pigs and yaks are reservoirs of non-O157 STECs in China,[35,36] and we also isolated various non-O157 STEC strains from different raw meats collected in different areas in China on a large scale.[37] Some of these non-O157 STEC isolates from different resources showed potential to cause human diseases based on comparisons by serotypes, Stx subtypes, presence of virulence genes, and sequence types with human pathogenic strains.

4.5 GENOMICS

E. coli K-12 MG1655 was the first genome-sequenced *E. coli* strain, which was published in 1997 by the Blattner et al.[38] and signaled a change in our understanding of model organisms; this was followed by the sequencing of two O157:H7 isolates, EDL933,[39] and the Sakai isolate.[40] The sequencing of these two O157:H7 isolates and the associated comparative analysis provided significant insights into the evolution of *E. coli* in general and EHEC specifically: (1) the isolates of EHEC were closely related and the analysis could distinguish differences in each group and (2) a significant amount of diversity within and between the EHEC isolates existed, but there was >1 Mb of DNA that was unique in the EHEC isolates that was not present in the laboratory-adapted K-12 MG1655 isolate. This represented approximately 20% of the genome that was unique in the pathogen, providing ample opportunity for functional characterization. These seminal publications provided the spark that allowed others to start functionally characterizing these genes and the role they play in either survival or pathogenesis in the human model systems. They also allowed comparison of EHEC to other pathotypes of *E. coli*.

Additional interesting genomes of the O157:H7 were published by Kulasekara et al.;[41] they sequenced an isolate, TW14359, which was associated with HUS. An additional 70 kb of genetic material that was not present in either of the reference O157 isolates (EDL933 and Sakai) were identified, which encoded additional putative Type III secreted effectors and a gene for an anaerobic nitric oxide reductase, *norV*. It was suggested that the *norV* gene could be used as a marker for increased virulence leading to HUS; however, screening of large numbers of isolates suggested that the *norV* gene was associated with the loss of stx_1, but the direct impact on pathogenesis was not confirmed. This publication is important in that it attempts to meld comparative genomics with disease presentation and gene presence or absence. This can be considered the beginning of the field we now know as genomic epidemiology.

The complete genome of *E. coli* O157:H7 strain Xuzhou21, which causing a major outbreak in China in 1999, revealed a novel conjugative plasmid pO157_Sal. The plasmid pO157_Sal were present absolutely in Chinese outbreak strains. Global transcriptional and phenotypic analyses demonstrated that the plasmid pO157_Sal affects the transcription of the chromosomal and pO157 plasmid genes and contributes to the enhanced ability to resist stress, suggesting that pO157_Sal plays an important role in regulating global gene expression and affects the virulence and adaptation of *E. coli* O157:H7 Xuzhou21.[42]

With the decrease in the cost of sequencing, an increasing number of EHEC isolates associated with disease within an outbreak setting have been sequenced and analyzed. Several studies have attempted to associate genetic changes with increases in virulence or markers that laboratories could use as identifiers of outbreak strains.[43] Also in 2009, the first non-O157 EHEC genomes including isolates from serotypes O26, O111, and O103 were published by Ogura et al.[44] These genomes were compared to the Sakai genome, demonstrating that the EHEC genomes were routinely larger than the MG1655 genome and contained a diverse array of phage and integrative elements, sometimes associated with the virulence gene catalog. Most importantly, it demonstrated how different serogroups and lineages of *E. coli* could evolve into EHEC isolates and demonstrated that there was a genomic core among similar isolates by comparing the gene content of 345 orthologous genes in all *E. coli* isolates that were sequenced. Meanwhile, the study implied that not all features of the genomes were the same.

4.6 PATHOGENESIS

E. coli O157:H7 infections may be acquired by direct contact with an infected person, domestic or wild animals, but the most frequent route is via consumption of contaminated food and water. Following ingestion, *E. coli* O157:H7 must survive the low acidity of the stomach. The ability to survive in the extremely low pH of the stomach increases the chance of bacteria to colonize the intestines and cause infection. Acid resistance is associated with a lowering of the infectious dose of enteric pathogens. The low infectious dose is one of the best known characteristics of *E. coli* O157:H7, making this bacterium highly infectious.[45]

E. coli O157:H7 are characterized for the presence of specific sets of virulence genes, including those encoding Stxs, intimin, enterohemolysin, and other fimbrial or afimbrial adhesins. The production of Stxs is the main virulence feature of STEC associated with the development of HUS but cannot be solely responsible for full pathogenicity. STEC associated with severe human disease is usually capable of colonizing the intestinal mucosa and possesses mobile genetic elements carrying virulence genes, such as phages, plasmids, transposons, and pathogenicity islands.

4.6.1 *SHIGA TOXIN*

Stx is derived from the dysentery bacillus *Shigella dysenteriae*, which was first described by Kiyoshi Shiga in 1898. Since then, numerous Stxs have been characterized. Stx family consist of a group of structurally and functionally closely related proteins.[46] The Stx family can be categorized into two types, Stx type 1 (Stx1) and Stx type 2 (Stx2). Stx1 is nearly identical to the *S. dysenteriae* toxin, differing by a single amino acid. The polyclonal antisera raised against Stx1 do not neutralize Stx2, which has about 60% sequence homology to Stx1. A number of Stxs variants are identified in STEC; Stx1 consists of three subtypes (Stx1a, Stx1c, and Stx1d), whereas Stx2 has seven subtypes (Stx2a, Stx2b, Stx2c, Stx2d, Stx2e, Stx2f, and Stx2g). STEC can carry a single variant, either Stx1 or Stx2, both Stx1 andStx2, or a combination of different Stx subtypes. Both Stx1- and Stx2-containing STEC can lead to HUS; however, Stx2a is a more potent toxin than Stx1.[47]

Stxs are produced via a single operon containing at minimum the *stxA* and *stxB* genes, and at least one promoter. The Stx operons are encoded on lambdoid prophages that are integrated into the chromosome. Stx1 and Stx2 are AB5 toxins, which are composed of a single copy of the A subunit and five copies of B subunit (Fig. 4.2). The B moiety consisted of five identical B subunits binds to globotriaosylceramide (Gb3) receptor present on host microvascular endothelial cell surfaces. The expression of Gb3 is high in renal glomerular endothelial cells of humans allowing the binding of Stxs followed by endocytosis of the toxin. The toxin is transported to the Golgi apparatus and endoplasmic reticulum. The N-terminal A1 domain subunit is cleaved from the C-terminal A2 domain by a protease. A disulfide bond is also reduced leading to full release of the A1 subunit, and the A1 subunit enters the cytosol via chaperone-mediated transfer. The A1 subunit is an N-glycosidase that removes a particular adenosine from the 3' region of the 28S rRNA of the 60S ribosomal subunit leading to the inhibition of protein synthesis. The renal glomerular lesions associated with HUS are due to damage of the endothelial cells. In addition to inhibiting protein synthesis, Stxs can trigger apoptosis in epithelial, endothelial, myeloid, and lymphoid cells in vitro. Apoptosis is a form of cell death resulting from the activation of intracellular signaling pathways. Apoptotic cells display characteristic morphological changes such as cell shrinkage, cytoplasmic vacuolation, chromatin condensation, nuclear fragmentation and cell blebbing to produce apoptotic bodies. Stxs may also indirectly induce apoptosis by inhibition of normal cell cycling, induction of proinflammatory and apoptosis-related gene expression, and attracting and activating cells capable of expressing apoptosis-inducing factors.[48]

Stxs are AB5 toxins. The B moiety consisted of five identical B subunits binds to Gb3 receptor; The A1 subunit is an N-glycosidase that is cleaved from the C-terminal A2 domain by a protease.

The production and release of Stxs are known to be linked to the promoters and associated genes of the lambdoid prophage genomes. Different environmental triggers, such as low environmental iron, environmental stressors, and particularly antibiotics that target DNA synthesis can induce Stxs expression. Stx prophages can become lytic during bacterial stress and is released from lysed bacterial cells during the lytic cycle of the phage. The use of antibiotics to treat *E. coli* O157:H7 infection has become contentious due to the stimulation of the lytic cycle and concomitant toxin release through the bacterial SOS response.

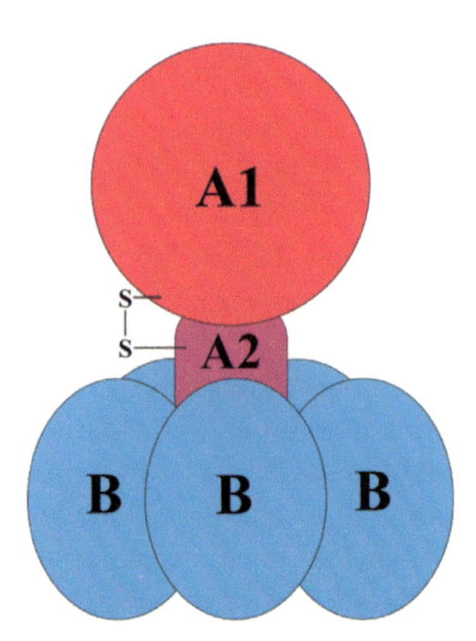

FIGURE 4.2 Schematic diagram of the Shiga toxin structure.

4.6.2 THE LOCUS OF ENTEROCYTE EFFACEMENT (LEE)

E. coli O157:H7 is able to produce attaching and effacing (A/E) lesions on intestinal epithelial cells. A/E lesions are characterized by intimate bacterial attachment to the apical surface of enterocytes, cytoskeletal rearrangements beneath adherent bacteria, and destruction of proximal microvilli. EPEC is the prototype organism causing A/E histopathology; however, other pathogens, including *E. coli* O157:H7, *Escherichia albertii* and *Citrobacter rodentium* are also members of the A/E pathogen family. The capacity for A/E lesion formation is encoded on the locus of enterocyte effacement (LEE), which encodes a Type III secretion system (T3SS). T3SSs are key virulence factors of Gram-negative enteric pathogens and serve to inject bacterial proteins directly into host cells.[49]

The LEE of *E. coli* O157:H7 is about 43 kb in size and composed of at least 41 different genes organized in five polycistronic operons, LEE1 through LEE5, two biocistronic operons, and four monocistronic entities. LEE1 to LEE3 operons encode a T3SS that exports effecter molecules, including (1) effecter molecules; (2) an adhesion called intimin and its translocated receptor, Tir; and (3) several secreted proteins (Esp). Recently, non-LEE encoded effectors (*nle*) have also been identified and shown to influence EHEC survival and colonization.

4.6.3 PO157 PLASMID

Plasmid-encoded genes are required for full pathogenesis in many enteropathogenic bacteria including *Shigella*, *Yersinia*, *Salmonella*, and *E. coli* species. In addition to Stxs and the LEE, which both are chromosomally encoded, all clinical isolates of *E. coli* O157:H7 possess a 60 MDa putative virulence plasmid named pO157. The pO157 is a nonconjugative F-like plasmid with a range size from 92 to 104 kb. This plasmid is also present in other but not all Stx-producing *E. coli* strains isolated from humans. The complete sequence of pO157 in two different outbreaks O157:H7 isolates EDL933 and Sakai has been published.[50,51] The pO157 shows a dynamic structure and includes different mobile genetic elements such as transposons, prophages, insertion sequences (IS), and parts of other plasmids. The complete sequence of pO157 reveals 100 open reading frames (ORFs).[50] Many proteins encoded on this plasmid are presumably involved in the pathogenesis of *E. coli* O157:H7 infections, including a hemolysin (*ehxA*),[52] a catalaseperoxidase (*katP*),[53] a type II secretion system apparatus (*etp*),[54] a serine protease (*espP*),[55] a putative adhesin (*toxB*),[56] a zinc metalloprotease (*stcE*),[57] etc.

Hemolysin was the first described virulence factor of pO157.[58,59] The hemolysin operon (*ehxCABD*) may be foreign in origin because it has a different G+C% and codon usage than the surrounding genetic contents. A 3.4 kb fragment encodes genes required for hemolysin synthesis and transport, and this region has been used as a diagnostic probe for *E. coli* O157:H7 and often EHEC isolates. Several studies showed that hemolysin is highly conserved among different serotypes of EHEC such as O157:H7, O111:H8, and O8:H19, but it is unknown if these have identical biological activities.[60]

A gene for a catalase-peroxidase activity (*katP*) was identified from pO157.[53] This gene is 2.2 kb in size and is highly homologous to the bacterial bifunctional catalase-peroxidase. The *katP* gene was found in all *E. coli* O157:H7 strains but not in EPEC, ETEC, EIEC, and EAEC strains. The KatP enzyme activity of *E. coli* O157:H7 was shown in both cytoplasm and periplasm fractions. This enzyme may help O157:H7 to colonize host intestines by reducing oxidative stress and using the by-product oxygen in diminished or deprived oxygen conditions of the host intestine.

The pO157 encodes 13 ORFs named *etpC* to *etpO*, which show high similarities to type two secretion system (T2SS) of Gram-negative bacteria.[54] These genes are located adjacent to the hemolysin locus. Similar to the *katP* gene, *etp* genes were also found in all *E. coli* O157:H7 strains, some in non-O157 EHEC strains, and not found in EPEC, ETEC, EIEC, and EAEC strains.

EspP is the pO157-encoded type V secretion system (T5SS) secreted serine protease and is known to cleave pepsin A and human coagulation factor V.[55] Dzivaet al.[61] reported that EspP influences the intestinal colonization of calves and adherence to bovine primary intestinal epithelial cells. Moreover, degradation of human coagulation factor V via EspP could contribute to the mucosal hemorrhage observed in HC patients.

The *toxB* gene is encoded on a sequence 9.5 kb in size, and its predicted product shows 20% similarity with toxin B of *Clostridium difficile*.[51] Studies showed that ToxB contributes to the adherence of *E. coli* O157:H7 to Caco-2 cells through the increased secretion of T3SS.[56]

A metalloprotease, StcE, is encoded on pO157 and specifically cleaves the C1 esterase inhibitor.[57] The C1 esterase inhibitor is a host regulator of multiple proteolytic cascades related to inflammation pathways, such as the classical complement, intrinsic coagulation, and contact activation. StcE is secreted through T2SS encoded on pO157 and is regulated by the LEE-encoded regulator (*ler*).[18,62] Grys et al.[63] demonstrated that StcE can contribute to intimate adherence of *E. coli* O157:H7 to Hep2 cells *in vitro*. The *stcE* gene was found in all *E. coli* O157:H7, in some EPEC serotype O55:H7 strains and it is not found in other DEC. Besides the genes described above, there are other genes reside on pO157 that associated with the potential pathogenesis.[64]

After the first report that pO157 was required for the expression of fimbriae and adhesion to epithelial cells, several studies reported conflicting results on the role of pO157 in adherence to epithelial cells.[65] Other investigators have reported that loss of this plasmid either decreased adhesion, enhanced adhesion, or had no effect on adhesion. Despite the role of this plasmid in the pathogenesis of EHEC is unknown, it is in fact widely distributed among human EHEC isolates.

4.6.4 OTHER POTENTIAL VIRULENCE FACTORS

In addition to the three major virulence factors of *E. coli* O157:H7 described above, other chromosomal and plasmid virulence genes also contribute in EHEC pathogenesis. Iha is an iron regulated gene A homolog adhesion that contributes to attachment of LEE-positive and LEE-negative strains.[66,67] Other virulence factor involves hemolysin encoded by *hlyA* and flagellar H7 gene encoded by *fliCh7*. [68] Nle effectors are proteins encoded on prophage elements throughout the genome [69] and are involved in altering host cell response. Other virulence genes, encoding cytotoxins (*subA*), adhesins (*saa*), etc., also contribute in STEC pathogenesis by allowing attachment and colonization of human epithelium.[4,70]

4.7 IDENTIFICATION AND DETECTION

There are many challenges in the isolation of EHEC from complex sample matrices. While clinical specimens may have high numbers of the pathogen present, animal feces, environmental and food samples may contain very low numbers of EHEC. In most matrices, there may be very high levels of background micro-flora and natural inhibitors which interfere with isolation and subsequent detection of the pathogen. Additionally, in some occasions, EHEC may be present in an injured or stressed condition as a result of food processing stresses (pH, temperature, preservatives, etc.) or the presence of antibiotics in clinical specimens. Unless an enrichment step is built into the protocol these cells may not be recovered, resulting in false negative results. These problems result in a necessary tradeoff between the need to incorporate antibiotics and other inhibitory agents into the enrichment broth and agar to enhance selectivity and the potential inhibition of stressed cells by these selective agents.

Classic detection methods are based on specific enrichment, often coupled with immunomagnetic separation (IMS) system, specific selective media, and different immunoassays. Molecular detection methods, based on DNA probes and PCR, are used to detect the virulence genes and the specific genes associated to the serogroups.

4.7.1 ENRICHMENT AND INOCULATING ON SELECTIVE MEDIA

A liquid enrichment step may be important for isolation of EHEC from all sample matrices in order to increase numbers of the pathogen to a detectable level and to allow recovery of injured cells. Two of the most successful enrichment media used for *E. coli* O157 and other EHEC serogroups are tryptone soy broth (TSB) and *E. coli* broth (ECB) with or without modifications to their original formulation. Modification to TSB media may include addition of bile salts and dipotassium phosphate while modified EC broth contains less bile salt. These basal media may also be supplemented with various selective agents like cefixime, potassium tellurite, and vancomycin to inhibit other Gram-negative flora.

For food and animal feeding stuffs there is an International Organization for Standardization (ISO) reference method (ISO 16654) for isolation of *E. coli* O157. This method advocates enrichment in modified TSB with novobiocin (mTSBn) at 41.5°C for an initial period of 6 h and then for a further incubation of 12–18 h at 37°C. At present, there is no standardized enrichment protocol for non-O157 STEC though a number of enrichment protocols have been reported which will allow for the isolation of a wide range of STEC serogroups or which have been optimized for recovery of selected serogroups.

The majority of commercial agars for EHEC still focus predominantly on the identification of *E. coli* O157. The inability of most *E. coli* O157 to ferment sorbitol is exploited in sorbitol McConkey agar (SMAC). Cexifime and potassium tellurite may be added to the SMAC (CT-SMAC) to increase selectivity in heavily contaminated samples. *E. coli* O157 generally produces colorless colonies when cultured on this media, thus distinguishing it from other STEC serogroups and other micro-flora. This is the media of choice in the ISO standard protocol (ISO 16654) for *E. coli* O157, together with a second appropriate selective agar.

Although non-sorbitol fermenting *E. coli* O157 predominates, sorbitol fermenting *E. coli* O157 (non-motile) have emerged as causes of HUS in Europe and Australia. These particular variants will not be readily identified on SMAC as they will produce colored colonies which will appear similar to the other micro-flora present. Additionally, the presence of potassium tellurite in CT-SMAC may actively inhibit the growth of sorbitol fermenting *E. coli* O157. Thus, alternative strategies are required for cultural detection of these strains, an example of which exploits the fact that in general *E. coli* O157 (including the sorbitol-fermenting strains) are β-glucuronidase (GUD) negative. Commercially developed examples of media which utilize this characteristic include 4-methylumbelliferyl-β-D-glucoronide (MUG) agar and 5-bromo-4-chloro-3-indolyl-β-D-glucoronide (BCIG) agar. Other commercial agars in use for recovery and identification of *E. coli* O157 are CHROMagar™O157, Rainbow Agar® O157, Fluorocult™ *E. coli* O157 Agar, Biosynth Culture Medium® O157:H7, etc.

Despite the identification method for O157:H7 serotype have been well-established, many challenges related to non-O157 STEC detection still remain, since these comprise a heterogeneous group of pathogens with different phenotypic features. In addition, the sensitivity of non-O157 strains to selective agents commonly used in enrichment media varies; therefore, determining which enrichment medium to use that allows growth of all non-O157 STEC strains has been a challenge. There are, however, some commercial chromogenic agars including Chromocult and Rainbow agar which differentiate between O157 and other selected STEC serogroups (O111, O26, O103, and O145) on a color basis. For STEC of other multiple serotypes, both low selective and high selective medium should be used conjunctively to facilitate the isolation of most STEC strains.

4.7.2 IMMUNOMAGNETIC SEPARATION

A number of isolation methods using antibodies specific to particular STEC serogroups are available. IMS recovers target cells from the enrichment broth using paramagnetic beads. These beads are coated with polyclonal antibodies specific for a particular EHEC serogroup. Beads coated with antibodies against serogroups O157, O26, O111, O103, and O145 are commercially available. The cell bead complex is recovered from the medium by the application of a magnetic field that causes the complex to be concentrated in a tube. The bulk of the medium is decanted off leaving the cell bead complex in the tube. The concentrated cell bead complex can then be examined by culturing onto solid media or by a rapid method such as PCR. IMS can be fully automated (e.g., BeadRetriever™; Dynal Biotech Ltd, Wirral, UK). In the gold standard cultural O157 method (ISO 16654), the inclusion of IMS is a prerequisite step before cultural isolation onto plating media. While there is no standardized protocol for other non-O157 STEC, IMS has been shown to be useful in recovery of specific serogroups from food and fecal samples.

4.7.3 IMMUNOLOGICAL METHODS

Immunological methods are widely used for the detection of STEC. The methods utilize specific poly- or monoclonal antibodies targeting surface antigens and thus detect specific STEC serogroups while others detect the toxins produced by STEC.

Most commercial kits target the O157 antigen expressed by bacteria in a mixed culture with few, if any, targeting other STEC serogroups. These are available in various formats including enzyme-linked immunosorbent assays (ELISA), reversed passive latex agglutination (RPLA) and automated systems combining IMS and ELISA. All commercial kits require prior enrichment of the target cells to reach detectable levels. These assays are generally less sensitive than traditional cultural techniques; however, they offer advantages in terms of rapidity, reduced labor costs and high volume throughput. All immunological techniques should be adequately standardized and controlled to reduce the possibility and impact of false positive and false negative results. This is more prevalent in samples with high levels of diverse background microflora. In all cases suspect positive samples should be subject to further confirmatory tests as the coating antibody may cross-react with other organisms. False positive results occur when the immunological material cross-reacts with a non-STEC organism (i.e., antibody is insufficiently specific). False negative results occur when the immunological material does not detect STEC cells when they are present in the sample under examination (i.e., antibody can find a target binding site on the STEC organism).

4.7.4 MOLECULAR DETECTION METHODS

The molecular definitions for EHEC and STEC are based on key virulence factors, that is, *stx* genes (stx_1, stx_2, or *stx* subtypes) as well as others such as *eae* (intimin) gene. Scheutz et al.[47] established a protocol for the subtyping of both stx_1 and stx_2 using a simple PCR-based method. Closely related strains of *E. coli* have *rfb* genes encoding different O antigens and this can be exploited to differentiate between different serogroups of STEC. Iguchi et al.[71] devised a basic set of PCR primers for the identification and classification of almost all known *E. coli* O serogroups (O1-O187). The PCR-based O-genotyping system reported there provides an accurate and reliable approach for O-genotyping *E. coli* isolates to much the same level as O serotyping. This *E. coli* O-genotyping PCR system might be a promising tool for subtyping of *E. coli* strains for epidemiological studies as well as for the surveillance of pathogenic *E. coli* during outbreaks.

With potentially lethal and widespread outbreaks of *E. coli* O157:H7, a large-scale and in-depth survey of genetic and architectural polymorphisms is a crucial prerequisite to obtain insights into the natural pathogenome evolution and the extent of bacterial disease virulence genotypes. Genetic heterogeneity among *E. coli* O157:H7 strains has been established by using a broad panel of targeted- and whole-genome-based typing assays to determine diversity and evolutionary relationships among isolates, such as pulsed-field gel electrophoresis (PFGE),[72] multi-locus sequence typing (MLST)[73] phage typing (PT),[74,75] multiple-locus variable-number tandem repeat analysis (MLVA),[76,77] microarrays,[78] single nucleotide polymorphism assays (SNP),[79,80] etc.

PFGE is widely used in research and epidemiological outbreak investigations to distinguish strains of *E. coli* O157:H7. The resulting PFGE profiles (DNA fingerprints) are analyzed and used for molecular comparison of isolates. Isolates with indistinguishable PFGE profiles are more likely to have a common source compared to isolates with distinct PFGE profiles.[81] The epidemiological value of subtyping *E. coli* O157:H7 isolates by PFGE has been well demonstrated in different studies.[72,82]

MLST was first developed for *Neisseria meningitidis* in 1998 to overcome the poor reproducibility between laboratories applying older molecular typing schemes.[83] It is a useful system for phylogenetic and epidemiological studies of *E. coli*. The principle behind the MLST scheme is to identify internal nucleotide sequences of approximately 400–600 base pairs in multiple housekeeping genes. There are currently three MLST schemes for *E. coli*, which are based on three different gene combinations.[84] The EcMLST database created by Thomas Whittam and hosted at Michigan State University (http://www.shigatox.net/ecmlst/cgi-bin/index) uses seven housekeeping genes (i.e., *aspC*, *clpX*, *fadD*, *icd*, *lysP*, *mdh*, and *uidA*)[85] The *E. coli* MLST database developed by Mark Acthman and hosted at the Warwick Medical School (http://mlst.warwick.ac.uk/mlst/dbs/Ecoli) employs another housekeeping genes except *icd* (i.e., *adk*, *fumC*, *gyrB*, *icd*, *mdh*, *purA*, and *recA*)[86] And the one devised by Sylvain Brisse and Erick Denamur and hosted at the Pasteur Institute (http://bigsdb.pasteur.fr/ecoli/ecoli.html) utilizes eight genes (i.e., *dinB*, *icdA*, *pabB*, *polB*, *putP*, *trpA*, *trpB*, and *uidA*).[87] Though these MLST methods are generally congruent, there are some differences for some strains. These MLST systems first determined that EHEC isolates are genetically highly clonal but could also be separated into multiple clades, suggesting that there are multiple evolutionary paths to generate a fully virulent EHEC isolate.

With the advent of high-throughput DNA sequencing techniques, and the decreased cost of sequencing, large numbers of *E. coli* O157:H7 isolates or non-O157 STEC isolates whole-genome sequencing are currently on way by several independent research teams.[88] Whole-genome mapping has been used for detailed genome comparisons to differentiate closely related STEC strains based on alterations in the chromosomal architectures (insertions, deletions, and rearrangements), which can help to reveal valid biological markers to trace evolution and also assists in genome assembly for molecular epidemiology outbreak investigations.

4.8 TREATMENT AND PREVENTION

4.8.1 TREATMENT

Currently no specific treatment is available for EHEC infections.[89] The use of conventional antibiotics exacerbates Stx-mediated cytotoxicity. In an epidemiology study conducted by the Centers for Disease Control and Prevention (CDC), patients treated with antibiotics for EHEC enteritis had a higher risk of developing HUS.[90] Additional studies support the contraindication of antibiotics in EHEC infection; children on antibiotic therapy for HC associated with EHEC had an increased chance of developing HUS.[91,92]

In light of the difficulties treating this pathogen with standard approaches, alternative modalities have been suggested to decrease the risk of systemic complications, consisting of supportive care and monitoring for the development of microangiopathic complications. Protection of renal function is achieved by intravenous volume expansion using isotonic crystalloids, especially if started early in the course of the disease.[5,93] It is important to remember that patients can seem to be deceptively well-hydrated while undergoing the initial phases of a severe prothrombotic process. Oral rehydration will not achieve the same beneficial effect on renal perfusion. Advanced supportive care in patients developing oligo-anuric renal failure includes blood transfusions, peritoneal dialysis, and plasma exchange, although recent retrospective reports have questioned the utility of plasma exchange.[94]Antiperistaltic agents have been shown to increase the risk of systemic complications and should be avoided.[95,96]

4.8.2 PREVENTION

As no effective treatment is available for EHEC infections, protecting yourself from being infected with this pathogen is important. *E. coli* O157 can be transmitted to humans through contaminated food and water, directly between persons, and through contact with animals or their environment. Given these, some prevention tips are suggested as follows.

4.8.2.1 BE CAREFUL WHEN IN CONTACT WITH ANIMALS

The reservoir of EHEC O157:H7 appears to be mainly cattle. In addition, other ruminants such as sheep, goats, deer are considered significant reservoirs, while other mammals (pigs, horses, rabbits, dogs, and cats), and birds (chickens and turkeys) have been occasionally found infected. One may get infected after close contact with the infected animals. Wash your hands thoroughly after contact with animals or their environments (at farms, petting zoos, fairs, even your own backyard).

4.8.2.2 AVOID RAW ANIMAL-DERIVED FOODS AND UNPASTEURIZED MILK.

E. coli O157:H7 is transmitted to humans primarily through consumption of contaminated foods, such as raw or undercooked ground meat products and raw milk. Fecal contamination of water and other foods, as well as cross-contamination during food preparation (with beef and other meat products, contaminated surfaces and kitchen utensils), will also lead to infection. Examples of foods implicated in outbreaks of *E. coli* O157:H7 include undercooked hamburgers, dried cured salami, unpasteurized fresh-pressed apple cider, yogurt, or cheese made from raw milk. Meat that has been needle-tenderized should be cooked to a temperature of at least 160°F/70°C. It is best to use a thermometer, as color is not a very reliable indicator of "doneness." An increasing number of outbreaks are associated with the consumption of fruits and vegetables (sprouts, spinach, lettuce, coleslaw, and salad) whereby contamination may be due to contact with feces from domestic or wild animals at some stage during cultivation or handling.

4.8.2.3 AVOID DRINKING UNPASTEURIZED WATER

EHEC has also been isolated from bodies of water (ponds and streams), wells, and water troughs, and has been found to survive for months in manure and water-trough sediments. A number of EHEC infections have been caused by contact with drinking-water and recreational water. Therefore, it is also important to protect such water areas, as well as drinking-water sources from animal waste.

4.8.2.4 AVOID PERSON-TO-PERSON INFECTION

Person-to-person contact is an important mode of transmission through the oral-fecal route. An asymptomatic carrier state has been reported, where individuals show no clinical signs of disease but are capable of infecting others. The duration of excretion of EHEC is about one week or less in adults, but can be longer in children. Visiting farms and other venues where the general public might come into direct contact with farm animals has also been identified as an important risk factor for EHEC infection.

STEC are foodborne pathogens, the most common means of spread, particularly in outbreak situations, is ingestion of contaminated food products. Thus, the prevention of infection requires control measures at all stages of the food chain, from agricultural production on the farm to processing, manufacturing and preparation of foods in both commercial establishments and household kitchens.

4.8.2.5 INDUSTRY

The number of cases of disease might be reduced by various mitigation strategies for ground beef (e.g., screening the animals pre-slaughter to reduce the introduction of large numbers of pathogens in the slaughtering environment). Good hygienic slaughtering practices reduce contamination of carcasses by feces, but do not guarantee the absence of EHEC from products.

Education in hygienic handling of foods for workers at farms, abattoirs and those involved in the food production is essential to keep microbiological contamination to a minimum. The only effective method of eliminating EHEC from foods is to introduce a bactericidal treatment, such as heating (e.g., cooking and pasteurization) or irradiation.

4.8.2.6 HOUSEHOLD

Preventive measures for *E. coli* O157:H7 infections are similar to those recommended for other foodborne diseases. Basic good food hygiene practice, as described in the WHO five keys to safer food, can prevent the transmission of pathogens responsible for many foodborne diseases, and also protect against foodborne diseases caused by EHEC.

Such recommendations should in all cases be implemented, especially "Cook thoroughly" so that the center of the food reaches at least 70°C. Make sure to wash fruits and vegetables carefully, especially if they are eaten raw. If possible, vegetables and fruits should be peeled. Vulnerable populations (e.g., small children, the elderly) should avoid the consumption of raw or undercooked meat products, raw milk, and products made from raw milk.

4.8.2.7 WASH YOUR HANDS REGULARLY

Regular hand washing, particularly before food preparation or consumption and after toilet contact, is highly recommended, especially for people who take care of small children, the elderly or immunocompromised individuals, as the bacteria can be passed from person-to-person, as well as through food, water, and direct contact with animals.

4.9 FUTURE PERSPECTIVE

Cattle are the main reservoir of *E. coli* O157:H7. Multiple colonization factors and cellular processes have been involved in the mechanism of *E. coli* O157:H7 adhesion to bovine intestinal epithelial. Thus, interventions that control cattle colonization, reduce pathogen concentration, or eliminate shedding, may be an effective attempt to reduce the transmission to human.

Public health surveillance of human *E. coli* O157:H7 infections can play an important role in implementing control measures. We cannot yet prevent or specifically treat HUS following *E. coli* O157:H7 infection. At present, there is also no effective vaccine against *E. coli* O157:H7. To find the biomarkers that discriminate related clinical syndromes, quantify the early onset or risk of HUS will be of greatest value.

Though the production of Stxs is the primary mechanism by which STEC damages humans, Stxs have become powerful delivery tools for the investigation of cellular processes and are being used for medical purposes, such as cancer therapy and imaging. To better understand the processes of Stxs entry into cells expressing Gb3, will provide us with a better therapy against *E. coli* O157:H7 infections, and may lead to better treatment of cancer.

KEYWORDS

- ***Escherichia coli***
- **hemorrhagic colitis**
- **O157:H7**
- **Shiga toxin**

REFERENCES

1. Croxen, M. A.; Law, R. J.; Scholz, R.; Keeney, K. M.; Wlodarska, M.; Finlay, B. B. Recent Advances in Understanding Enteric Pathogenic *Escherichia coli. Clin. Microbiol. Rev.* **2013,** *26* (4), 822–880.
2. Riley, L. W.; Remis, R. S.; Helgerson, S. D.; McGee, H. B.; Wells, J. G.; Davis, B. R., et al. Hemorrhagic Colitis Associated with a Rare *Escherichia coli* Serotype. *N. Engl. J. Med.* **1983,** *308* (12), 681–685.
3. Rangel, J. M.; Sparling, P. H.; Crowe, C.; Griffin, P. M.; Swerdlow, D. L. Epidemiology of *Escherichia coli* O157:H7 Outbreaks, United States, 1982–2002. *Emerg. Infect. Dis.* **2005,** *11* (4), 603–609.
4. Melton-Celsa, A.; Mohawk, K.; Teel, L.; O'Brien, A. Pathogenesis of Shiga-toxin Producing *Escherichia coli. Curr. Top. Microbiol. Immunol.* **2012,** *357,* 67–103.
5. Smith, J. L.; Fratamico, P. M.; Gunther, N. W., 4th. Shiga Toxin-producing *Escherichia coli. Adv. Appl. Microbiol.* **2014,** *86,* 145–197.
6. Davis, T. K.; McKee, R.; Schnadower, D.; Tarr, P. I. Treatment of Shiga Toxin-producing *Escherichia coli* Infections. *Infect. Dis. Clin. North Am.* **2013,** *27* (3), 577–597.
7. Mayer, C. L.; Leibowitz, C. S.; Kurosawa, S.; Stearns-Kurosawa, D. J. Shiga Toxins and the Pathophysiology of Hemolytic Uremic Syndrome in Humans and Animals. *Toxins (Basel).* **2012,** *4* (11), 1261–1287.
8. Marshall, J. K. Post-infectious Irritable Bowel Syndrome Following Water Contamination. *Kidney Int. Suppl.* **2009,** *112,* S42–S43.

9. Bavaro, M. F. *E. coli* O157:H7 and Other Toxigenic Strains: The Curse of Global Food Distribution. *Curr. Gastroenterol. Rep.* **2012,** *14* (4), 317–323.
10. Karch, H.; Tarr, P. I.; Bielaszewska, M. Enterohaemorrhagic *Escherichia coli* in Human Medicine. *Int. J. Med. Microbiol.* **2005,** *295* (6–7), 405–418.
11. Schuster, C. J.; Ellis, A. G.; Robertson, W. J.; Charron, D. F.; Aramini, J. J.; Marshall, B. J., et al. Infectious Disease Outbreaks Related to Drinking Water in Canada, 1974–2001. *Can. J. Public Health.* **2005,** *96* (4), 254–258.
12. Serna, A. 4th.; Boedeker, E. C. Pathogenesis and Treatment of Shiga Toxin-producing *Escherichia coli* Infections. *Curr. Opin. Gastroenterol.* **2008,** *24* (1), 38–47.
13. Meng, Q.; Xiong, Y.; Lan, R.; Ye, C.; Wang, T.; Qi, T., et al. SNP Genotyping of Enterohemorrhagic *Escherichia coli* O157:H7 Isolates from China and Genomic Identity of the 1999 Xuzhou Outbreak. *Infect. Genet. Evol.* **2013,** *16,* 275–281.
14. Xiong, Y.; Wang, P.; Lan, R.; Ye, C.; Wang, H.; Ren, J., et al. A Novel *Escherichia coli* O157:H7 Clone Causing a Major Hemolytic Uremic Syndrome Outbreak in China. *PLoS One* **2012,** *7* (4), e36144.
15. Pruimboom-Brees, I. M.; Morgan, T. W.; Ackermann, M. R.; Nystrom, E. D.; Samuel, J. E.; Cornick, N. A., et al. Cattle Lack Vascular Receptors for *Escherichia coli* O157:H7 Shiga Toxins. *Proc. Natl. Acad. Sci. USA* **2000,** *97* (19), 10325–10329.
16. Conedera, G.; Mattiazzi, E.; Russo, F.; Chiesa, E.; Scorzato, I.; Grandesso, S., et al. A Family Outbreak of *Escherichia coli* O157 Haemorrhagic Colitis Caused by Pork Meat Salami. *Epidemiol. Infect.* **2007,** *135* (2), 311–314.
17. Ferguson, D. D.; Scheftel, J.; Cronquist, A.; Smith, K.; Woo-Ming, A.; Anderson, E., et al. Temporally Distinct *Escherichia coli* 0157 Outbreaks Associated with Alfalfa Sprouts Linked to a Common Seed Source--Colorado and Minnesota, 2003. *Epidemiol. Infect.* **2005,** *133* (3), 439–447.
18. Grant, J.; Wendelboe, A. M.; Wendel, A.; Jepson, B.; Torres, P.; Smelser, C., et al. Spinach-associated *Escherichia coli* O157:H7 Outbreak, Utah and New Mexico, 2006. *Emerg. Infect. Dis.* **2008,** *14* (10), 1633–1636.
19. Neil, K. P.; Biggerstaff, G.; MacDonald, J. K.; Trees, E.; Medus, C.; Musser, K. A., et al. A Novel Vehicle for Transmission of *Escherichia coli* O157:H7 to Humans: Multistate Outbreak of *E. coli* O157:H7 Infections Associated with Consumption of Ready-to-bake Commercial Prepackaged Cookie Dough--United States, 2009. *Clin. Infect. Dis.* **2012,** *54* (4), 511–518.
20. Hunt, J. M. Shiga Toxin-producing *Escherichia coli* (STEC). *Clin. Lab. Med.* **2010,** *30* (1), 21–45.
21. Nielsen, S.; Frank, C.; Fruth, A.; Spode, A.; Prager, R.; Graff, A., et al. Desperately Seeking Diarrhea: Outbreak of Haemolytic Uraemic Syndrome Caused by Emerging Sorbitol-fermenting Shiga Toxin-producing *Escherichia coli* O157:H, Germany, 2009. *Zoonoses Public Health.* 2011, *58* (8), 567–572.
22. Watahiki, M.; Isobe, J.; Kimata, K.; Shima, T.; Kanatani, J.; Shimizu, M., et al. Characterization of Enterohemorrhagic *Escherichia coli* O111 and O157 Strains Isolated from Outbreak Patients in Japan. *J. Clin. Microbiol.* **2014,** *52* (8), 2757–2763.
23. Slayton, R. B.; Turabelidze, G.; Bennett, S. D.; Schwensohn, C. A.; Yaffee, A. Q.; Khan, F., et al. Outbreak of Shiga Toxin-producing *Escherichia coli* (STEC) O157:H7 Associated with Romaine Lettuce Consumption, 2011. *PLoS One* **2013,** *8* (2), e55300.

24. Nabae, K.; Takahashi, M.; Wakui, T.; Kamiya, H.; Nakashima, K.; Taniguchi, K., et al. A Shiga Toxin-producing *Escherichia coli* O157 Outbreak Associated with Consumption of Rice Cakes in 2011 in Japan. *Epidemiol. Infect.* **2013,** *141* (9), 1897–1904.
25. Marder, E. P.; Garman, K. N.; Ingram, L. A.; Dunn, J. R. Multistate Outbreak of *Escherichia coli* O157:H7 Associated with Bagged Salad. *Foodborne Pathog. Dis.* **2014,** *11* (8), 593–595.
26. Bayliss, L.; Carr, R.; Edeghere, O.; Knapper, E.; Nye, K.; Harvey, G., et al. School Outbreak of *Escherichia coli* O157 with High Levels of Transmission, Staffordshire, England, February 2012. *J. Public Health (Oxf)* **2015,** *38,* e247–e253.
27. Tabuchi, A.; Wakui, T.; Yahata, Y.; Yano, K.; Azuma, K.; Yamagishi, T., et al. A Large Outbreak of Enterohaemorrhagic *Escherichia coli* O157, Caused by Low-salt Pickled Napa Cabbage in Nursing Homes, Japan, 2012. *Western Pac. Surveill. Response J.* **2015,** *6* (2), 7–11.
28. Gaulin, C.; Ramsay, D.; Catford, A.; Bekal, S. *Escherichia coli* O157:H7 Outbreak Associated with the Consumption of Beef and Veal Tartares in the Province of Quebec, Canada, in 2013. *Foodborne Pathog. Dis.* **2015,** *12* (7), 612–618.
29. Rowell, S.; King, C.; Jenkins, C.; Dallman, T. J.; Decraene, V.; Lamden, K., et al. An Outbreak of Shiga Toxin-producing *Escherichia coli* Serogroup O157 Linked to a Lamb-feeding Event. *Epidemiol. Infect.* **2016,** *144,* 2494–2500.
30. Majowicz, S. E.; Scallan, E.; Jones-Bitton, A.; Sargeant, J. M.; Stapleton, J.; Angulo, F. J., et al. Global Incidence of Human Shiga Toxin-producing *Escherichia coli* Infections and Deaths: A Systematic Review and Knowledge Synthesis. *Foodborne Pathog. Dis.* **2014,** *11* (6), 447–455.
31. Coombes, B. K.; Wickham, M. E.; Mascarenhas, M.; Gruenheid, S.; Finlay, B. B.; Karmali, M. A. Molecular Analysis as an Aid to Assess the Public Health Risk of Non-O157 Shiga Toxin-producing *Escherichia coli* Strains. *Appl. Environ. microbiol.* **2008,** *74* (7), 2153–2160.
32. Karmali, M. A.; Petric, M.; Lim, C.; Fleming, P. C.; Arbus, G. S.; Lior, H. The Association between Idiopathic Hemolytic Uremic Syndrome and Infection by Verotoxin-producing *Escherichia coli. J. Infect. Dis.* **1985,** *151* (5), 775–782.
33. Tozzi, A. E.; Caprioli, A.; Minelli, F.; Gianviti, A.; De Petris, L.; Edefonti, A., et al. Shiga Toxin-producing *Escherichia coli* Infections Associated with Hemolytic Uremic Syndrome, Italy, 1988–2000. *Emerg. Infect. Dis.* **2003,** *9* (1), 106–108.
34. Brooks, J. T.; Sowers, E. G.; Wells, J. G.; Greene, K. D.; Griffin, P. M.; Hoekstra, R. M., et al. Non-O157 Shiga Toxin-producing *Escherichia coli* Infections in the United States, 1983–2002. *J. Infect. Dis.* **2005,** *192* (8), 1422–1429.
35. Bai, X.; Zhao, A.; Lan, R.; Xin, Y.; Xie, H.; Meng, Q., et al. Shiga Toxin-producing *Escherichia coli* in Yaks (*Bos grunniens*) from the Qinghai-Tibetan Plateau, China. *PLoS One* **2013,** *8* (6), e65537.
36. Meng, Q.; Bai, X.; Zhao, A.; Lan, R.; Du, H.; Wang, T., et al. Characterization of Shiga Toxin-producing *Escherichia coli* Isolated from Healthy Pigs in China. *BMC Microbiol.* **2014,** *14,* 5.
37. Bai, X.; Wang, H.; Xin, Y.; Wei, R.; Tang, X.; Zhao, A., et al. Prevalence and Characteristics of Shiga Toxin-producing *Escherichia coli* Isolated from Retail Raw Meats in China. *Int. J. FoodMicrobiol.* **2015,** *200,* 31–38.

38. Blattner, F. R.; Plunkett, G., 3rd.; Bloch, C. A.; Perna, N. T.; Burland, V.; Riley, M., et al. The Complete Genome Sequence of *Escherichia coli* K-12. *Science* **1997,** *277* (5331), 1453–1462.
39. Perna, N. T.; Plunkett, G. 3rd.; Burland, V.; Mau, B.; Glasner, J. D.; Rose, D. J., et al. Genome Sequence of Enterohaemorrhagic *Escherichia coli* O157:H7. *Nature* **2001,** *409* (6819), 529–533.
40. Hayashi, T.; Makino, K.; Ohnishi, M.; Kurokawa, K.; Ishii, K.; Yokoyama, K., et al. Complete Genome Sequence of Enterohemorrhagic *Escherichia coli* O157:H7 and Genomic Comparison with a Laboratory Strain K-12. *DNA Res.* **2001,** *8* (1), 11–22.
41. Kulasekara, B. R.; Jacobs, M.; Zhou, Y.; Wu, Z.; Sims, E.; Saenphimmachak, C., et al. Analysis of the Genome of the *Escherichia coli* O157:H7 2006 Spinach-associated Outbreak Isolate Indicates Candidate Genes That may Enhance Virulence. *Infect. Immun.* **2009,** *77* (9), 3713–3721.
42. Zhao, H.; Chen, C.; Xiong, Y.; Xu, X.; Lan, R.; Wang, H., et al. Global Transcriptional and Phenotypic Analyses of *Escherichia coli* O157:H7 Strain Xuzhou21 and Its pO157_Sal Cured Mutant. *PLoS One* **2013,** *8* (5), e65466.
43. Eppinger, M.; Mammel, M. K.; Leclerc, J. E.; Ravel, J.; Cebula, T. A. Genomic Anatomy of *Escherichia coli* O157:H7 Outbreaks. *Proc. Natl. Acad. Sci. USA* **2011,** *108* (50), 20142–20147.
44. Ogura, Y.; Ooka, T.; Iguchi, A.; Toh, H.; Asadulghani, M.; Oshima, K., et al. Comparative Genomics Reveal the Mechanism of the Parallel Evolution of O157 and Non-O157 Enterohemorrhagic *Escherichia coli. Proc. Natl. Acad. Sci. USA* **2009,** *106* (42), 17939–17944.
45. Hong, W.; Wu, Y. E.; Fu, X.; Chang, Z. Chaperone-dependent Mechanisms for Acid Resistance in Enteric Bacteria. *Trends Microbiol.* **2012,** *20* (7), 328–335.
46. Bergan, J.; DyveLingelem, A. B.; Simm, R.; Skotland, T.; Sandvig, K. Shiga Toxins. *Toxicon.* **2012,** *60* (6), 1085–1107.
47. Scheutz, F.; Teel, L. D.; Beutin, L.; Pierard, D.; Buvens, G.; Karch, H., et al. Multicenter Evaluation of a Sequence-based Protocol for Subtyping Shiga Toxins and Standardizing Stx Nomenclature. *J. Clin. Microbiol.* **2012,** *50* (9), 2951–2963.
48. Tesh, V. L. Induction of Apoptosis by Shiga Toxins. *Future Microbiol.* **2010,** *5* (3), 431–453.
49. Stevens, M. P.; Frankel, G. M.; The Locus of Enterocyte Effacement and Associated Virulence Factors of Enterohemorrhagic *Escherichia coli. Microbiol. Spectr.* **2014,** *2* (4), EHEC-0007–2013.
50. Burland, V.; Shao, Y.; Perna, N. T.; Plunkett, G.; Sofia, H. J.; Blattner, F. R. The Complete DNA Sequence and Analysis of the Large Virulence Plasmid of *Escherichia coli* O157:H7. *Nucleic. Acids Res.* **1998,** *26* (18), 4196–4204.
51. Makino, K.; Ishii, K.; Yasunaga, T.; Hattori, M.; Yokoyama, K.; Yutsudo, C. H., et al. Complete Nucleotide Sequences of 93-kb and 3.3-kb Plasmids of an Enterohemorrhagic *Escherichia coli* O157:H7 Derived from Sakai Outbreak. *DNA Res.* **1998,** *5* (1), 1–9.
52. Schmidt, H.; Karch, H.; Beutin, L. The Large-sized Plasmids of Enterohemorrhagic *Escherichia coli* O157 Strains Encode Hemolysins Which are Presumably Members of the *E. coli* Alpha-hemolysin Family. *FEMS Microbiol. Lett.* **1994,** *117* (2), 189–196.

53. Brunder, W.; Schmidt, H.; Karch, H. KatP, A Novel Catalase-peroxidase Encoded by the Large Plasmid of Enterohaemorrhagic *Escherichia coli* O157:H7. *Microbiology* **1996,** *142* (Pt 11), 3305–3315.
54. Schmidt, H.; Henkel, B.; Karch, H. A Gene Cluster Closely Related to Type II Secretion Pathway Operons of Gram-negative Bacteria is Located on the Large Plasmid of Enterohemorrhagic *Escherichia coli* O157 Strains. *FEMS Microbiol. Lett.* **1997,** *148* (2), 265–272.
55. Brunder, W.; Schmidt, H.; Karch, H. EspP, A Novel Extracellular Serine Protease of Enterohaemorrhagic *Escherichia coli* O157:H7 Cleaves Human Coagulation Factor V. *Mol. Microbiol.* **1997,** *24* (4), 767–778.
56. Tatsuno, I.; Horie, M.; Abe, H.; Miki, T.; Makino, K.; Shinagawa, H., et al. *toxB* Gene on pO157 of Enterohemorrhagic *Escherichia coli* O157:H7 is Required for Full Epithelial Cell Adherence Phenotype. *Infect. Immun.* **2001,** *69* (11), 6660–6669.
57. Lathem, W. W.; Grys, T. E.; Witowski, S. E.; Torres, A. G.; Kaper, J. B.; Tarr, P. I., et al. StcE, A Metalloprotease Secreted by *Escherichia coli* O157:H7, Specifically Cleaves C1 Esterase Inhibitor. *Mol. Microbiol.* **2002,** *45* (2), 277–288.
58. Bauer, M. E.; Welch, R. A. Characterization of an RTX Toxin from Enterohemorrhagic *Escherichia coli* O157:H7. *Infect Immun.* **1996,** *64* (1), 167–175.
59. Schmidt, H.; Beutin, L.; Karch, H. Molecular Analysis of the Plasmid-encoded Hemolysin of *Escherichia coli* O157:H7 Strain EDL 933. *Infect. Immun.* **1995,** *63* (3), 1055–1061.
60. Brashears, M. M.; Galyean, M. L.; Loneragan, G. H.; Mann, J. E.; Killinger-Mann, K. Prevalence of *Escherichia coli* O157:H7 and Performance by Beef Feedlot Cattle Given Lactobacillus Direct-fed Microbials. *J. Food Prot.* **2003,** *66* (5), 748–754.
61. Dziva, F.; Mahajan, A.; Cameron, P.; Currie, C.; McKendrick, I. J.; Wallis, T. S., et al. EspP, A Type V-secreted Serine Protease of Enterohaemorrhagic *Escherichia coli* O157:H7, Influences Intestinal Colonization of Calves and Adherence to Bovine Primary Intestinal Epithelial Cells. *FEMS Microbiol. Lett.* **2007,** *271* (2), 258–264.
62. Elliott, S. J.; Sperandio, V.; Giron, J. A.; Shin, S.; Mellies, J. L.; Wainwright, L., et al. The Locus of Enterocyte Effacement (LEE)-encoded Regulator Controls Expression of Both LEE- and Non-LEE-encoded Virulence Factors in Enteropathogenic and Enterohemorrhagic *Escherichia coli. Infect. Immun.* **2000,** *68* (11), 6115–6126.
63. Grys, T. E.; Siegel, M. B.; Lathem, W. W.; Welch, R. A. The StcE Protease Contributes to Intimate Adherence of Enterohemorrhagic *Escherichia coli* O157:H7 to Host Cells. *Infect Immun.* **2005,** *73* (3), 1295–1303.
64. Lim, J. Y.; Yoon, J.; Hovde, C. J. A Brief Overview of *Escherichia coli* O157:H7 and Its Plasmid O157. *J. Microbiol. Biotechnol.* **2010,** *20* (1), 5–14.
65. Yoon, J. W.; Hovde, C. J. All Blood, No Stool: Enterohemorrhagic *Escherichia coli* O157:H7 Infection. *J. Vet. Sci.* **2008,** *9* (3), 219–231.
66. Schmidt, H.; Zhang, W. L.; Hemmrich, U.; Jelacic, S.; Brunder, W.; Tarr, P. I., et al. Identification and Characterization of a Novel Genomic Island Integrated at *selC* in Locus of Enterocyte Effacement-negative, Shiga Toxin-producing *Escherichia coli. Infect. Immun.* **2001,** *69* (11), 6863–6873.
67. Tarr, P. I.; Bilge, S. S.; Vary, J. C., Jr.; Jelacic, S.; Habeeb, R. L.; Ward, T. R., et al. Iha: A Novel *Escherichia coli* O157:H7 Adherence-conferring Molecule Encoded on

a Recently Acquired Chromosomal Island of Conserved Structure. *Infect. Immun.* **2000,** *68* (3), 1400–1407.

68. Saxena, T.; Kaushik, P.; Krishna Mohan M. Prevalence of *E. coli* O157:H7 in Water Sources: An Overview on Associated Diseases, Outbreaks and Detection Methods. *Diagn. Microbiol. Infect. Dis.* **2015,** *82* (3), 249–264.
69. Holmes, A.; LindestamArlehamn, C. S.; Wang, D.; Mitchell, T. J.; Evans, T. J.; Roe, A. J. Expression and Regulation of the *Escherichia coli* O157:H7 Effector Proteins NleH1 and NleH2. *PLoS One* **2012,** *7* (3), e33408.
70. Bolton, D. J. Verocytotoxigenic (Shiga Toxin-producing) *Escherichia coli:* Virulence Factors and Pathogenicity in the Farm to Fork Paradigm. *Foodborne Pathog. Dis.* **2011,** *8* (3), 357–365.
71. Iguchi, A.; Iyoda, S.; Seto, K.; Morita-Ishihara, T.; Scheutz, F.; Ohnishi, M., et al. *Escherichia coli* O-Genotyping PCR: A Comprehensive and Practical Platform for Molecular O Serogrouping. *J. Clin. Microbiol.* **2015,** *53* (8), 2427–2432.
72. Gerner-Smidt, P.; Kincaid, J.; Kubota, K.; Hise, K.; Hunter, S. B.; Fair, M. A., et al. Molecular Surveillance of Shiga Toxigenic *Escherichia coli* O157 by PulseNet USA. *J. Food Prot.* **2005,** *68* (9), 1926–1931.
73. Afset, J. E.; Anderssen, E.; Bruant, G.; Harel, J.; Wieler, L.; Bergh, K. Phylogenetic Backgrounds and Virulence Profiles of Atypical Enteropathogenic *Escherichia coli* Strains from a Case-control Study Using Multilocus Sequence Typing and DNA Microarray Analysis. *J. Clin. Microbiol.* **2008,** *46* (7), 2280–2290.
74. Ahmed, R.; Bopp, C.; Borczyk, A.; Kasatiya, S. Phage-typing Scheme for *Escherichia coli* O157:H7. *J. Infect. Dis.* **1987,** *155* (4), 806–809.
75. Bono, J. L.; Smith, T. P.; Keen, J. E.; Harhay, G. P.; McDaneld, T. G.; Mandrell, R. E., et al. Phylogeny of Shiga Toxin-producing *Escherichia coli* O157 Isolated from Cattle and Clinically Ill Humans. *Mol. Biol. Evol.* **2012,** *29* (8), 2047–2062.
76. Keys, C.; Kemper, S.; Keim, P. Highly Diverse Variable Number Tandem Repeat Loci in the *E. coli* O157:H7 and O55:H7 Genomes for High-resolution Molecular Typing. *J. Appl. Microbiol.* **2005,** *98* (4), 928–940.
77. Noller, A. C.; McEllistrem, M. C.; Pacheco, A. G.; Boxrud, D. J.; Harrison, L. H. Multilocus Variable-number Tandem Repeat Analysis Distinguishes Outbreak and Sporadic *Escherichia coli* O157:H7 Isolates. *J. Clin. Microbiol.* **2003,** *41* (12), 5389–5397.
78. Jackson, S. A.; Mammel, M. K.; Patel, I. R.; Mays, T.; Albert, T. J.; LeClerc, J. E., et al. Interrogating Genomic Diversity of *E. coli* O157:H7 Using DNA Tiling Arrays. *Forensic Sci. Int.* **2007,** *168* (2–3), 183–199.
79. Zhang, W.; Qi, W.; Albert, T. J.; Motiwala, A. S.; Alland, D.; Hyytia-Trees, E. K., et al. Probing Genomic Diversity and Evolution of *Escherichia coli* O157 by Single Nucleotide Polymorphisms. *Genome Res.* **2006,** *16* (6), 757–767.
80. Manning, S. D.; Motiwala, A. S.; Springman, A. C.; Qi, W.; Lacher, D. W.; Ouellette, L. M., et al. Variation in Virulence among Clades of *Escherichia coli* O157:H7 Associated with Disease Outbreaks. *Proc. Natl. Acad. Sci. USA* **2008,** *105* (12), 4868–4873.
81. Jaros, P.; Dufour, M.; Gilpin, B.; Freeman, M. M.; Ribot, E. M. PFGE for Shiga Toxin-producing *Escherichia coli* O157:H7 (STEC O157) and Non-O157 STEC. *Methods Mol. Biol.* **2015,** *1301,* 171–189.

82. Terajima, J.; Izumiya, H.; Iyoda, S.; Mitobe, J.; Miura, M.; Watanabe, H. Effectiveness of Pulsed-field Gel Electrophoresis for the Early Detection of Diffuse Outbreaks due to Shiga Toxin-producing *Escherichia coli* in Japan. *Foodborne Pathog. Dis.* **2006,** *3* (1), 68–73.
83. Maiden, M. C.; Bygraves, J. A.; Feil, E.; Morelli, G.; Russell, J. E.; Urwin, R., et al. Multilocus Sequence Typing: A Portable Approach to the Identification of Clones within Populations of Pathogenic Microorganisms. *Proc. Natl. Acad. Sci. USA* **1998,** *95* (6), 3140–3145.
84. Clermont, O.; Gordon, D.; Denamur, E. Guide to the Various Phylogenetic Classification Schemes for *Escherichia coli* and the Correspondence among Schemes. *Microbiology* **2015,** *161* (Pt 5), 980–988.
85. Reid, S. D.; Herbelin, C. J.; Bumbaugh, A. C.; Selander, R. K.; Whittam, T. S. Parallel Evolution of Virulence in Pathogenic *Escherichia coli. Nature* **2000,** *406* (6791), 64–67.
86. Wirth, T.; Falush, D.; Lan, R.; Colles, F.; Mensa, P.; Wieler, L. H., et al. Sex and Virulence in *Escherichia coli:* An Evolutionary Perspective. *Mol. Microbiol.* **2006,** *60* (5), 1136–1151.
87. Jaureguy, F.; Landraud, L.; Passet, V.; Diancourt, L.; Frapy, E.; Guigon, G., et al. Phylogenetic and Genomic Diversity of Human Bacteremic *Escherichia coli* Strains. *BMC Genomics.* **2008,** *9,* 560.
88. Hazen, T. H.; Sahl, J. W.; Fraser, C. M.; Donnenberg, M. S.; Scheutz, F.; Rasko, D. A. Refining the Pathovar Paradigm Via Phylogenomics of the Attaching and Effacing *Escherichia coli. Proc. Natl. Acad. Sci. USA* **2013,** *110* (31), 12810–12815.
89. Goldwater, P. N.; Bettelheim, K. A. Treatment of Enterohemorrhagic *Escherichia coli* (EHEC) Infection and Hemolytic Uremic Syndrome (HUS). *BMC Med.* **2012,***10,* 12.
90. Slutsker, L.; Ries, A. A.; Maloney, K.; Wells, J. G.; Greene, K. D.; Griffin, P. M. A Nationwide Case-control Study of *Escherichia coli* O157:H7 Infection in the United States. *J. Infect. Dis.* **1998,** *177* (4), 962–966.
91. Safdar, N.; Said, A.; Gangnon, R. E.; Maki, D. G. Risk of Hemolytic Uremic Syndrome after Antibiotic Treatment of *Escherichia coli* O157:H7 Enteritis: A Meta-Analysis. *JAMA* **2002,** *288* (8), 996–1001.
92. Tarr, P. I.; Gordon, C. A.; Chandler, W. L. Shiga-toxin-producing *Escherichia coli* and Haemolytic Uraemic Syndrome. *Lancet* **2005,** *365* (9464), 1073–1086.
93. Rahal, E. A.; Kazzi, N.; Nassar, F. J.; Matar, G. M. *Escherichia coli* O157:H7-clinical Aspects and Novel Treatment Approaches. *Front. Cell. Infect. Microbiol.* **2012,** *2,* 138.
94. Hauswaldt, S.; Nitschke, M.; Sayk, F.; Solbach, W.; Knobloch, J. K. Lessons Learned from Outbreaks of Shiga Toxin Producing *Escherichia coli. Curr. Infect. Dis. Rep.* **2013,** *15* (1), 4–9.
95. Thielman, N. M.; Guerrant, R. L. Clinical Practice. Acute Infectious Diarrhea. *N. Engl. J. Med.* **2004,** *350* (1), 38–47.
96. Nelson, J. M.; Griffin, P. M.; Jones, T. F.; Smith, K. E.; Scallan, E. Antimicrobial and Antimotility Agent Use in Persons with Shiga Toxin-producing *Escherichia coli* O157 Infection in FoodNet Sites. *Clin. Infect. Dis.* **2011,** *52* (9), 1130–1132.

CHAPTER 5

DIARRHEA AND ENTEROTOXIGENIC *ESCHERICHIA COLI*

XIN-HE LAI[1,2,#*], LONG-FEI ZHAO[1#], and YAN QIAN[2*]

[1]*College of Life Sciences, Key Laboratory of Plant-Microbe Interactions of Henan, Shangqiu Normal University, 55 Pingyuanzhong Road, Shangqiu, Henan, PR China 476000*

[2]*Department of Pediatrics, the First Affiliated Hospital of Wenzhou Medical University, New Campus at Nanbaixiang, Ouhai District, Wenzhou, China 325000*

[*]*Corresponding author. E-mail: laixinhe@yahoo.com; qianyan11@126.com*

CONTENTS

Abstract 124
5.1 Introduction 124
5.2 Biological Characteristics 125
5.3 Diseases 125
5.4 Epidemiology 126
5.5 Genomics 131
5.6 Pathogenesis 132
5.7 Identification and Detection 139
5.8 Treatment and Prevention 140
5.9 Future Perspectives 144
Keywords 144
References 144

[#]Equal contribution.

ABSTRACT

Enterotoxigenic *Escherichia coli* (ETEC) is one of the causal agents of diarrhea, most frequently seen in children in the developing world. This chapter reviews how people get into contact with ETEC, its transmission pathway in the body, and the up-to-date understanding of ETEC's pathogenic mechanism(s) to make human sick and cause diarrhea. The introduction of these basic knowledges hopefully will serve the purpose steering people from the risk of contracting ETEC-related diseases and edging responsible institutions to further improve access, quantity, and quality of public health facilities to curtail the threat of ETEC.

5.1 INTRODUCTION

Defined by the World Health Organization (WHO), the passage of three or looser than normal stools in the preceding 24 h period, diarrhea is a common illness affecting millions of people each year, a leading cause of death for children under 5 years of age in less developed countries and can be caused by many pathogens. Enterotoxigenic *Escherichia coli* (ETEC) is one of the major diarrheal agents. Although the estimated numbers can fluctuate between studies, ETEC infect millions of people and cause many deaths, either reported or unreported. According to a report on global causes of diarrheal disease, ETEC infection is estimated to having caused 42,000 deaths in 2011 in children <5 years of age.[1] A lot can be done for individuals to prevent diarrhea and for public/private organizations to improve the infrastructure of the drinking/irrigation system and monitor food safety.

An oversimplified version of how ETEC cause illness is that ETEC get into contact with human beings through ingestion/contacting contaminated items (mostly water and/or food), colonize the small intestine via adhesins, such as the colonization factors (CFs),[2] secrete heat-labile (LT) and/or heat-stable (ST) enterotoxins,[3,4] and ultimately cause diarrhea.[5–7]

Because of its host tropism, ETEC strains from animals seldom infect human. Due to the diversity of ETEC strains, mainly different genetic makeups, it is extremely important to keep in mind when reading findings from different labs that it really needs caution to generalize findings from one ETEC strain and extrapolate to others. This chapter summarizes recent advances in researches of human ETEC, gives a touch on its general aspects, points out some key findings, and focuses on its pathogenesis and prevention.

5.2 BIOLOGICAL CHARACTERISTICS

As a part of the normal intestinal microflora, *E. coli* is just one of the 35,000 bacterial species in the human gut,[8] and the majority of it is non-pathogenic. Only a fraction of *E. coli* is pathogenic, and causes intestinal and extra-intestinal infections of different nature. The main manifestation of intestinal *E. coli* infection is diarrhea that can be caused by different pathotypes, collectively named as diarrheagenic *E. coli* (DEC), including diffusely adherent *E. coli* (DAEC), enteroaggregative *E. coli* (EAEC), enteroinvasive *E. coli* (EIEC), enteropathogenic *E. coli* (EPEC), enterotoxigenic *E. coli* (ETEC), and Shiga toxin-producing *E. coli* (STEC). Unlike the other pathotypes being able to recognize from their typical colony color on a specific culture plate, it is hard to tell apart of ETEC from the normal *E. coli* in the routine culture unless utilizing molecular technique to discriminate them as mentioned briefly in the Detection Section of this chapter.

A typical ETEC bacterium is about the size of 0.25–1.0 μm wide and 2.0 μm long. Although 37°C is the optimal growth temperature, ETEC can manage to survive for months in fresh and seawater.[9,10]

5.3 DISEASES

First of all, ETEC has a fairly high infectious dose in the range of 10^8–10^{10} cells as estimated in volunteer feeding studies.[11,12] It might be different if caused by those newly hybrid strains with addition of other virulence factors from other pathotype(s).

Although there are lots of observation about how ETEC-infected patients are doing, sometimes it is difficult to ascertain that some symptoms are surely caused by ETEC considering that the possibility of co-infection of different pathogens is as high as to 40%[13] and some of the detections is done for a single agent only, in this case, ETEC. In this sense, if a volunteer challenge study is double blinded, the symptoms should be less objective and more trustworthy, already knowing the cause of the symptoms.

From the list of varied symptoms observed from volunteer/challenge studies, one will realize the need for systematic evaluation of the disease severity based on evidence, as one study points out[14] and many have experienced. The most common symptoms are diarrhea (77%), abdominal pain (67%), nausea (63%), abdominal cramping (53%), and fever (23%), which is done in fasting adult volunteers orally taking 1×10^6–1×10^9 colony forming units (CFUs) of the wild-type ETEC strain TW10598.[15] The rate of diarrhea can range between 44[12] and 88%,[16] and the severity and duration can be different depending on the strategy (whether fasting and bicarbonate are utilized), dose and ETEC strain[17]challenged. Other symptoms can also be presented varying from loss of appetite, intestinal gurgling, headaches, malaise, and fatigue.[17] For instance, B7A strain caused diarrhea with shorter incubation time and milder degree in 75% of the volunteers at a dose of 10^{10} CFU while H10407 could do starting from 10^8 CFU.[17] High fecal concentrations of H10407 strain are found to be associated with symptomatic ETEC infections.[18]

Symptoms can also be related. For example, vomiting and nausea is found to have the strongest correlation in a study among 264 subjects receiving seven ETEC strains at doses from 1×10^5 to 1×10^{10} cfu.[14] As a comparison, some of the common seen clinical symptoms from recent outbreaks are listed in Table 5.1.

5.4 EPIDEMIOLOGY

Like many other pathogens, there are mainly two ways to collect epidemiological information for ETEC. One is through outbreak reports to get a glimpse of the impact from ETEC and another is surveillance, which is time- and resource-intensive.

5.4.1 OUTBREAK

The scale of ETEC-related outbreaks can be small or big,[19–22] depending on the virulence of the involved strain(s) and the affected population size of the contamination. Most outbreaks are associated with water supply contamination[23,24] and food poisoning at parties or in isolated spaces like cruise ship,[23] prison,[25] ward,[26] and sometimes school[19,21] due to contaminated supply of food/water. Foods associated with outbreaks are mainly those uncooked/undercooked or/and contaminated ones (also mentioned in the Food Safety Section), including, but not limited to (lettuce) salad[27,28] sushi,[29] tuna paste[30] vegetables like kimchi,[19,21] carrot[31] chives,[20] and parsley.[32]

TABLE 5.1 Some Recent ETEC Outbreaks Worldwide Since 2000.

Series	Month/ year	Location	Setting	Source	Number of cases	O:K:H serotype	ST/LT	Odds ratio	Diarrhea-cramp-vomiting-nausea-fever%	Ref.
1	07/13	Korea	School	Kimchi	1189	O6	ST, LT	1.7	100%-91%-?-24%-?	[19]
2	12/12	Norway	Buffet	Chives	300	O78	LT	9.07	90%-90%-18%-50%-31%	[20]
3	09/12	Korea	School	Kimchi	1642	O169	/	4.52	22-94%-58-89%-?-?-13-58%	[21]
4	09/12	Japan	Eatery	Food handler?	102	O169:H41	ST	/	100%-68.6%-5.9%-25.5%-36.3%	[36]
5	01/10	Denmark	/	Lettuce	260	O6:K15	STh, LT	6.2		[28]
6	10/07	China	Hospital	Care worker	20	/	/	/	Diarrhea	[26]
7	11/06	Denmark	Canteen	Basil	217	O92:H-	STh, STp	2.6	95%-75%-31%-16.6%-62%	[27]
8	08/05	Japan	Prison	Kimuchi	401	O6:H16	ST, LT	/	95%-80%	[25]
9	09/04	USA	Cafeteria	Salad	240	O6:H16	ST, LT	4.5	100%-97.3%-21.1%-60.4%-29.5%	[33]
10	08/04	USA	Eatery	Sushi	113	O6:H16	ST, LT	7.2	100%-94%-22%-62%-40%	[29]
11	08/03	USA	Catering	Coleslaw	36	O169:H14	ST	4.4	94%-74%-?-?-31%	[37]
12	08/00	India	NICU	Cook	16	/	LT	/	Watery diarrhea	[38]
13	01/00	India	Hospital	/	809	O1, O146	LT	/	Secretory diarrhea	[39]

/, ?: not available

Some O serotype ETEC strains are more often associated with outbreaks, such as O6.[19,24,28,29,32–34] For reasons unknown, some serotypes (e.g., O169:H41) become more associated with outbreaks after certain time point. One American study reported that before 1996 only 1 out of 21 confirmed outbreaks was caused by O169:H41. After that, 10 out of 16 outbreaks have been caused by this serotype.[38]

According to our unexhausted search result from PubMed, O6:H16 had caused the following outbreaks in the past, 2 at the same day in Chiba City near Tokyo from consuming meals from the same supplier,[25] one each in Denmark from consuming lettuce[27] from the park's water supply at Crater Lake,[24] one of the 3 serotypes involved in a large (~100 persons) foodborne outbreak with a longer duration of illness in Illinois, USA[22] and one from eating sushi in Nevada.[29] Table 5.1 summarizes some of the features of the recent outbreaks[19–21,25–29,33,35–37,39] since 2000.

5.4.2 SURVEILLANCE

Surveillance tools, sample processing and data interpretation need to be improved all the time as it is the case with the Global Enteric Multicenter Study. When the samples were reanalyzed with quantitative real-time polymerase chain reaction (PCR) instead of reverse-transcriptase PCR as utilized previously, the new approach significantly improved the quality of diarrhea characterization both at the population level and case level.[40] With more and more detection tools available and affordable, surveillance has been feasible in some resource-limited regions. One can get some information like the endemic serotypes and affected population but it is difficult to do data comparison due to variations of the techniques used. Most epidemiological studies can only reach a conclusion that a certain pathogen is associated with diarrhea outbreaks in humans due to limitations like technical, moral, and other reasons. Some early studies had focused on a specific pathogen in their investigation and therefore a negative result simply helps to rule out the possible involvement of that specific pathogen and does not tell much about the etiology of the diarrhea. With about 17% chance, a person can be colonized by multiple pathogenic ETEC isolates.[41] It adds a further layer of complexity for result interpretation when more than one type of microbes are isolated from diarrheal cases as seen in some of the studies.[42] Even a pathogen is suspected as the causal agent of diarrhea, it fails to fulfill the criteria of Koch's postulates.

ETEC might be underestimated. ETEC has been found to disappear rapidly from the stools,[43,44] so strain isolation might be a methodology that would get a rate under the real number, in a way underestimating the epidemiological situation of ETEC. Although the current strategy focuses more on symptomatic patients, people carrying ETEC without symptom(s) draws increasing attention and may also be epidemiologically interesting. Moreover, with the emergence of hybrid strains as described below, it is difficult to decide which part of their hybrid nature should the epidemiological data to put with.

The diversity of ETEC further complicates surveillance in many ways. Already with over hundred O and H antigen types of complex combinations, the addition to ETEC of serotypes traditionally related to another pathotype and the emergence of hybrid strains of different nature further complicate strain typing, diagnostics and therefore the surveillance. ETEC O157 strains (SvETEC, *E. coli* O157:K88 734/3) belong to the ST23 complex[45] phylogenetically distinct from EHEC O157. So far there have been reports of hybrid strains of EPEC/ETEC,[46] and STEC/ETEC isolated from animal in Burkina Faso[47] and from human in Finland[48] and USA,[49,50] which harbor *stx* and *est* genes but lack the CFs.

The ranking position of ETEC as a diarrheal pathogen can be different depending on the region, season, and other factors. A sentinel surveillance in Minnesota during 2000–2006 found ETEC (1.9%) as the second (after *Campylobacter*) most common bacterial pathogen causing diarrhea.[51] According to data between 2004 and 2010 from 16 communities in Northwestern Ecuador, ETEC is the dominant pathogen and the regional prevalence of ETEC is relatively stable, ranging from 8 to 37 infections/1000 persons.[52] In a multicenter study of stool samples from Europe, ETEC are among the six most commonly detected pathogens,[53] similar to its position in the Mayo study.[54] Even in a high-income nation like Israel, ETEC ranks second (5%) among diarrheagenic *E. coli* in severe diarrheal samples.[55] Surprisingly, among the five diarrheagenic *E. coli,* ETEC is the primary pathogen of adult diarrhea in China, according to a recent survey.[42]

The epidemiology of ETEC might be different too in travelers due to difference in the age, immune status, geographical distribution, and so on. ETEC (together with EAEC, EPEC, and *Campylobacter) were found to be more commonly associated with* diarrheal *cases with* symptoms *in Finland* when comparing 382 travelers for their pre- and post-travel stools.[44] It is slightly different in a Spanish study where ETEC came second after

Shigella as the most common pathogen for traveler's diarrhea using a commercially available multiplex PCR examining 185 stool specimens obtained from 174 patients.[56]

A few temporal sectional studies have summed up the episode and mortality situation of ETEC in 2010,[57–59] which should provide readers a big picture and some clue how is ETEC retrospectively around the world at a period time past, as detailed in Table 5.2.

TABLE 5.2 Estimation of Diarrheal Incidence and Mortality Caused by ETEC in 2010.

Area	Episode	Mortality	Ref.
World	233 million	73,041	[58]
World	/	120,800	[59]
Africa	5 million	42,973	[57]
South Asia	28.7 million	45,713	[57]
AFRO	2970/100,000	4/100,000	[58]
AMRO	3816/100,000	0.1/100,000	[58]
EMRO	14,201/100,000	1/100,000	[58]
EURO	15/100,000	0	[58]
SEARO	3016/100,000	2/100,000	[58]
WPRO	1559/100,000	0.1/100,000	[58]
USA	13.3 million	/	[58]
	Among 10 foodborne diarrheal pathogens##		
61 nation+	210,362	0	[58]
	15/100,000	0	[58]
133 nations^	240,672,835	73,857	[58]
	4,404/100,000	1/100,000	[58]

/: not available;

AFRO African Region: 46 countries with 847,062,195 population;

AMRO American Region: 35 countries with 937,213,444 population;

EMRO Europe Mediterranean Region: 22 countries with 589,012,258 population;

EURO European Region: 53 countries with 898,751,529 population;

SEARO South East Asia Region: 11 countries with 1,789,987,553 population;

WPRO Western Pacific Region: 27 countries with 1,818,793,708 population;

##10 foodborne diarrheal pathogens: *Campylobacter*, EPEC, ETEC, norovirus, *Salmonella, Shigella, Cryptosporidium, Giardia, Entamoeba histolytica*, and others;

+61 nation: in AMRO subregion A (low mortality), EURO and WPRO subregion A;

^133 nations: the remaining nations with middle/high mortality.

5.5 GENOMICS

Although most ETEC virulence factors are encoded on plasmids, chromosome and some mobile elements like phages do play a role.

5.5.1 PLASMIDS

In contrary to most pathogens, it was known for a long time that the ETEC plasmids seem to play bigger role(s) than the chromosome in the pathogenesis. A single ETEC strain (E24377A) can carry as many as six plasmids of sizes varying from 5 to 79 Kb.[7,60,61] Both the enterotoxin genes and the plasmid from strain H10407 were sequenced.[62,63] So far, plasmids have been found to encode most of the ETEC pathogenesis-related factors, including CfaD-dependent expression extracytoplasmic protein (CexE), colonization factor/1 (CFA/I), EAEC heat-stable enterotoxin (EAST), ETEC autotransporter A (EatA), ETEC two-partner operon (EtpBAC), heat-labile enterotoxin (LT), heat-stable enterotoxin (ST) 1a (ST-P) and ST 1b (ST-H). Plasmid p7v carries two copies of *sta* with identical sequences.[49]

5.5.2 CHROMOSOME

ETEC is not any *E. coli* lineage plus plasmid(s) harboring CFs and/or toxins. Availability of more genome sequences and all the emerging omics provide us a panoramic view of the biological events and make meaningful comparative genomic analyses feasible. The genomes of distinct and globally dominant ETEC clonal groups are compared, finding that ETEC strains share a genomic core that is more conserved than non-ETEC *E. coli* genomes.[64] A recent study analyzed 362 representative strains from the ETEC collection of more than 3500 isolates and found that they formed into more than 20 discrete, identifiable and phylogenetically-related lineages (L1-L21) with consistent plasmid-encoded virulence profiles.[65] The chromosome and plasmid combinations can optimize ETEC fitness and transmissibility.[64] The gene repertoire of the "prototypical" ETEC strain H10407 is significantly different from those of some other known ETEC isolates[62] and no significant homologues of labile enterotoxin output A (*leoA)*, enterotoxigenic invasion locus C (*tibC)*, *tibA*, or enterotoxigenic invasion locus A (*tia*) in strain H10407 were detected in either *E. coli* strains E24377A or B7A.[66] Highly virulent hybrid pathotype strains that possess molecular markers associated with more than one *E. coli* pathotype have been reported. It is known that label-free quantitative MS/MS can improve microbial genome annotation quality[67] and there would be other improvements in the future in this fast moving field.

5.6 PATHOGENESIS

Studies of ETEC pathogenesis have been centered on the CFs and LT/ST toxins encoded on the plasmids, the following description is the short version how ETEC infection leads to diarrhea. Upon consumption via the contaminated water and/or food, ETEC must first survive the digestion in the acidic environment of the stomach, and pass into the small intestine. With the help of CFA and/or other adhesin(s), ETEC bacteria colonize the epithelial tissue, produce/secrete toxins and other virulence factor(s), and ultimately cause diarrhea. Discoveries of new adhesins and other potential virulence factors encoded on the chromosome prompt one to think ETEC pathogenesis is far more than the above mentioned.

5.6.1 BINDING AND COLONIZING THE INTESTINE

Before colonizing the intestine ETEC must be able to attach to epithelial cells, which involves flagella, CFs, and possibly other molecules. CFs (also called adhesins), either pili (fimbriae) or pilus-related molecules, can be fimbrial (pilus), fibrillar, helical, or afimbrial according to its morphology and are encoded on plasmid(s). More than 25 CFs have been described in the literature,[2] and CFA/I, coli *s*urface antigens (CS1-CS3, CS5, and CS6) are the most common ones.[68]

ETEC expressing the flexible CFA/I pili or similar structure(s) have the capacity to withstand harsh motion without breaking[69] which allows the pathogen to resist intestinal peristalsis, interact with their cognate receptors, attach/adhere to the intestinal epithelium, and hence begin to establish colonization in the intestine. ETEC can also use *E. coli* type-II partner A (EtpA) as a molecular bridge, attach it to the distal tips of the long peritrichous flagellar structures (up to 10 μm long), and make the initial contact with the epithelial surface.[70] Other outer membrane molecules like entero-attaching and effacing H (EaeH)[71] can mediate more intimate interactions with the host cell, bringing ETEC into close proximity to surface receptors for the delivery of LT and ST. Production of LT can also aid ETEC in adherence to the ileum.[72]

The functional efficacy of CFs can be affected by at least two factors, its expression level on the bacterial surface and CFs' intrinsic affinity to host cell surface receptors. Intestinal colonization by ETEC strains expressing CS6 involves interactions with fibronectin and sulpholipid.[73,74] Two CS6 subtypes show differential cellular binding.[75] The amino acid substitutions at positions 37 in coli surface structure A (CssA)II and 97 in CssBII in AIIBII (a CS6 subtype) decrease its surface expression as well as its affinity for mucin although total amount of CS6 is similar to AIBI (the wild type CS6), which contributes to significant reduction in the colonization efficiency of this mutant.[76]

5.6.2 LT AND ST TOXINS

Although carrying *eltAB,* not all LT+ strains produce the same amount of toxin, and also they can produce but not necessarily secrete the same amount of LT, most probably due to differential regulation. An example of this case is that LT2 strains (a LT variant) produce 5-fold more mature toxin over another lineage LT1 ones.[77] Of note is that LT-hyperproducing strains have been isolated from asymptomatic children with equal frequency as from diarrheal patients.[78]

This is the abbreviated scenario how ETEC bacteria from outside get into our body and travel all the way through the intestine. Before entry of the human host, no LT is produced under an outside environment of low osmolarity and ambient temperature. Upon entry into the gastrointestinal tract, ETEC first needs to survive the acidic stress in the gastric lumen, and soon to adapt to an alkaline environment in the duodenum and small intestine. There, the overall effect of higher temperature,[79] basic pH,[80] the presence of bicarbonate, bile[81] and free glucose[41] induces expression of virulence factors including the production/secretion of LT and ST after the inhibition of cAMP receptor protein (CRP) and H-NS is de-repressed.[82,83] Active secretion of LT by at least three different mechanisms, including the general secretory pathway (GSP) encoded by *gsp* genes[84] an ETEC-specific secretion pathway encoded by the *leo* gene[85] and the shedding of LT-containing membrane vesicles.[86] Down in the large intestine, the increasing amount of CRP and short-chain fatty acids represses *eltAB* expression[87] before its exit of the human body.

Following is an oversimplified view of how toxins get into targeted cells and exert their diarrheagenic effect based on *in vitro* and *in vivo* data generated all these years. LT-I is 78% identical to cholera toxin (CT) gene (*ctxAB*) in the overall nucleotide sequence although the B subunits of LT-II (LT-IIa, LT-IIb, and LT-IIc) are not significantly homologous to those of LT-I.[88,89] Considering the similarity between LT and CT in genetics, structure and more importantly functions, some results are borrowed from *V. cholerae* studies to make the gap(s) connected where appropriate. Once LTA and LTB subunits reach the periplasm, they quickly assemble there forming its AB_5 structure as holotoxins[90] in a DsbA-dependent manner[91] and cross the outer membrane through a type II secretion system.[88] LT can bind to the bacterial LPS on the extracellular surface[88] and the host cell receptor. After its B subunits bind to monosialoganglioside G_{M1} centered in Caveolae or other receptor(s) on the host cell,[92] LT is internalized and retrograde transports through the Golgi apparatus to the endoplasmic reticulum. The disulfide bond between A1 and A2 is reduced once B subunit contacts the cell surface, A1 subunit retrotranslocates into the cytosol of the cell after A2 (the helical *C*-terminal) is cleaved off.[93] Inside the cell, the catalytically active A1 interacts allosterically with ADP ribosylating factors and ADP-ribosylates G_{sa} protein in the arginine residue of its α subunit, leading to inhibition of the GTPase activity of $G_{s\alpha}$ and constitutively activation of adenylate cyclase. The intracellular cAMP increase alone or together with cGMP induced by ST leads to protein kinase A (PKA) activation and chloride channel (e.g., CFTR) phosphorylation, which increases Cl^- secretion across the cytoplasmic membrane into the intestinal lumen and reduces Na^+ absorption,[94] and diarrhea occurs.[94] Furthermore, the A subunit can stimulate intestinal secretion through prostaglandin E2 (PGE_2) and 5-hydroxytryptamine (5-HT).[95]

ST works similarly to LT with telling differences. There are at least two types of ST, STI (STh and STp) and STII (or STb). H10407 strain is an example that one strain can carry two subtypes of ST (STh and STp).[96] Following is the current understanding how ST alone induces diarrhea. During translocation across the inner membrane, the Sec-dependent signal sequence of pre-pro-peptide (72 aa) of STI is removed, and in the periplasm, the propeptide is catalyzed by DsbA to form disulfide bonds before secretion through TolC. The toxin matures, as an 18 or 19 amino acid peptide after removal of the proregion, and is secreted across the outer membrane. Binding of ST to the extracellular receptor domain of guanylate cyclase C (GC-C) causes conformational change of GC-C, which activates the guanylate cyclase domain, and increases the intracellular Cgmp.[97] In turn, cGMP directly activates protein kinase G II and PKA, indirectly activates PKA through inhibition of phosphodiesterase 3. Protein kinase G II phosphorylates cystic fibrosis transmembrane conductance regulator (CFTR)[98,99] which increases salt and water secretion and at the same time inhibits Na^+ absorption through an apical Na/H exchanger.[97] The end result of stimulating these three different signaling pathways ultimately causes the watery diarrhea. In some cells, STI can release calcium from the intracellular pool, which translocates protein kinase C (PKC) to the host cell membrane to activate guanylate cyclase, another pathway to increase cGMP.[100]

Different from STI, STII (48 aa) binds to sulfatide on the host cells, leading to an increase in Ca^{2+}. The influx of Ca^{2+} activates calmodulin-dependent protein kinase II, and opens an intestinal ion channel. The calcium increase can also produce PGE_2 and 5-HT that transport water and electrolyte out of intestinal cells[101] resulting in the secretory diarrhea associated with ETEC infection in animals. Similar to LT, STII may also activate PKC and activate CFTR, the main mediator of chloride secretion.

There is a possibility that LT and ST can be synergistic in intestinal secretion of fluid for LT/ST strain[102] or mixed infection with LT and ST strains.[41] The role of ST and LT in ETEC infection is conversely illustrated in a challenge study using mutant strains (PTL002 (Δ*aroC* Δ*ompR*) and PTL003 (Δ*aroC* Δ*ompC* Δ*ompF*) of E1392/75-2A) that are both LT and ST negative. They were well tolerated (with almost no diarrhea, vomit, or cramps) at doses from 5×10^7 to 5×10^9 CFU.[103]

There are exceptions for some ETEC strains. For H10407, LT secretion depends on LeoA[104] but only about 3% strains carry the *leoA* gene.[66] In H10407, LT secretion is also associated with outer membrane vesicles.[105]

5.6.3 OTHER VIRULENCE FACTORS

There is no doubt that CFs are essential for initiation of ETEC infection but there are strains without any known CF that are able to cause diarrhea,[77] suggesting something that can function in replacing CFs remains to be determined. Some exploring effort along this line is summarized below and there will be more such factors to be revealed in the future to complete the picture of ETEC pathogenesis.

EtpA, a two-partner secretion (TPS) exoprotein adhesin of ETEC, mimics and interacts with highly conserved regions of flagellin, the major subunit of flagella, and that these interactions are critical for adherence and intestinal colonization of the small intestine[106,107] and toxin delivery to epithelial cells.[107] YghJ too can enhance LT delivery through its metalloprotease activity and degradation of MUC2.[108] Another protease EatA can modulate ETEC adhesion and enhance LT delivery by degrading EtpA.[109] *Tia* is encoded within a pathogenicity island[105] and binds to heparin–sulfate proteoglycans on eukaryotic cell surfaces.[110] The unglycosylated TibA promotes aggregation of a bacterial population and the formation of a biofilm, and the glycosylated TibA directs the bacteria to bind to a specific receptor on epithelial cells.[111] Although similar to ST, the role of EAST1 in ETEC diarrhea remains to be determined.[112]

5.6.4 REGULATION OF VIRULENCE FACTOR EXPRESSION

The expression, secretion, and delivery of virulence factors in ETEC are under strict regulation in a temporal and spatial manner, which is still not completely understood. ETEC respond to a wide array of environmental cues to survive and cause infection, and following is a briefing what is currently known. Mainly due to different genetic makeup, the influence of a single factor on a lot of bacterial activities might be strain specific.

The report utilizing elegant Hadamard matrix designs has tested the effect of components (up to 20) under 24–36 ETEC growth conditions on CFA/I expression, LT secretion and growth,[113] exemplifying how to systemically evaluate the different roles each of the complex and interdependent cues plays when examining a complex biological process.

5.6.5 BILE SALTS

CS1 and CS3 expression is inhibited by the presence of bile salts.[114] CS14 expression in SP1 medium needs bile salts but addition of bile salts into M9 medium is not enough for CS14 expression.[113] The glyco-conjugated primary bile salt sodium glycocholate hydrate (NaGCH) induces phenotypic expression of CS5 in a dose-dependent manner and causes a 100-fold up-regulation of CS5 mRNA levels.[115]

5.6.6 QUORUM-SENSING

Many of the known or suspected virulence factors of ETEC are upregulated in the presence of bile compared to LB medium alone by a density-dependent manner, suggesting involvement of quorum sensing.[113] Not much research activities are in this direction. Upon inactivation of *luxS* and *pfs*, the ability of ETEC to produce the quorum-sensing signal, and induce auto-inducer 2 activity and biofilm formation is significantly reduced.[116] A significant increase in motility, F4 and LT expression, is observed in the ETEC culture supplemented with pre-conditioned medium (PCM) and epinephrine.[117] Expression of *estA*, *eltA*, CS1, and CS3 is significantly greater under all PCM conditions compared to the baseline condition.[114]

5.6.7 GLUCOSE/cAMP

Glucose, one of the components most studied, significantly upregulates *eltA* expression in strain H10407.[114,118] 0.2% glucose or cAMP induces CFA/I surface expression.[113] The presence of glucose in the growth medium does not affect expression of either enterotoxin genes, *estA,* or *eltA,* but upregulates the expression of *astA* and *eatA* in strain E24377A.[114] CRP is a negative regulator for the transcription of *eltAB*, and a positive regulator of the secretion of the LT.[80] The concentration of glucose used in oral rehydration treatment solution (ORT) is ~10-fold higher than its required to repress *estA2* expression,[82] so there should be sufficient glucose present in ORT to down regulate toxin expression.[82] After testing several concentrations, only media containing 0.25% glucose resulted in increased adherence and cAMP levels of LT^+ ETEC on IPEC-J2 cells.[119]

5.6.8 IRON

Whereas iron induces CFA/I surface expression with promoting bacterial growth in one study,[113] it is found that addition of 0.05 mM ferrous sulfate to growth media decreases CFA/I antigen and fimbrial production by the CFA/I-positive ETEC strain H-10407 as measured by quantitative enzyme -linked immunosorbent assay (ELISA) and hemagglutination assay.[120] The discrepancy between the studies remains to be resolved. Under iron starvation, production of the CFA/I fimbriae is increased in the ETEC H10407 prototype strain.[121] In contrast, LT secretion is inhibited.[113]

5.6.9 OXYGEN

LT secretion depends on the integrity of the respiratory chain and a functional GspD under anaerobic conditions,[122] more research should be done under physiological condition(s) close to the human body.

5.6.10 CELL-CONTACT

marA and *eatA* are upregulated upon attachment to Caco-2 cells while *eaeH* is upregulated at later time points.[123] In contrast, *cfaABC* was repressed following contact.

5.6.11 OTHER FACTORS

Zinc induces LT secretion[124] but inhibits CFA/I expression.[113] Inhibition of CFA/I expression by the zinc inhibitor 1,10-phenanthroline might be due to its nonspecific chelating of other metal cations.[113] ETGA induces bacterial growth and surface expression of the CFA/I.[113] Lincomycin induces LT secretion in strain H10407[125] but not in strain 258909–3.[113]

As pointed out in one report,[113] CFA/I surface expression responds to concentration changes of components in a complex way as they all (7 in total) interact to each other's changes with the exception of glucose. Glutamine, pH, and lincomycin each interacts with three other components while EGTA does only with porcine gastric mucin.[113] Understandably the expression/secretion of other factors might respond similarly in a complex system such as *in vivo* condition where all parameters could be different between individuals or the same individual under different time and various physical/psychological status.

5.6.12 PATHOGENIC MECHANISM(S)

Apart from the impact on cAMP, cGMP, PKC, PKA, and so on exerted by enterotoxins as described above, ETEC as a whole and its products can further modulate the signal transduction pathways and affect host cell survival.

LT can activate NF-κB following the activation of Rap1 (a Ras-like GTPase) independent of PKA.[126] In contrast, an uncharacterized factor secreted by H10407 strain can block NF-κB pathway[127] It remains unclear what is the overall effect of ETEC infection if the strain has both the activation and inhibition of NF-κB from LT and the unknown, respectively. Some other studies found that ETEC strains activate TLR4, MyD88, NF-κB, and MAPK pathways,[128,129] by phosphorylating IKKα, IKKβ, IκBα, and NF-κB subunit p65,[128] thus increasing production of inflammatory cytokines, such as IL-8 and IL-1β.[128]

ETEC infection can kill effector cells of important function. Strain H10407 was able to induce apoptosis in J774 macrophages,[130] and following studies incriminate its enterotoxin as the apoptogenic factor activating caspases[131],[132] ETEC can also induce autophagy in IPEC-1 cells by inhibiting mTOR through the AMPK and ERK1/2 signaling pathways.[133]

The following two studies provide a panoramic view of how ETEC infection profoundly affects cells specifically and human in general. Although, it is a transcriptome study of interaction of porcine intestinal epithelial cells (IPEC-J2) with ETEC of porcine origin, nevertheless providing insight to human study. ETEC infection of IPEC-J2 cells causes differential expression of 2443 unique genes, especially down-regulating the MAPK pathway and up-regulating the expression of important pro-inflammatory molecules including *IL-6*, *IL-8*, *TNF-α*, *CCL20*, and *CXCL2*.[134] Similar work in humans, apparently more complicated, is less conclusive.[135]

5.7 IDENTIFICATION AND DETECTION

Although serotyping is routine for strain characterization, it needs to isolate the strain(s) in colony form from a sample and the diagnosis therefore would be impossible if colony isolation is failed because ETEC was found to disappear rapidly from the stools.[43] As an alternative, a DNA microarray was developed for simultaneously detecting 19 commonest

O serogroup and enterotoxin genes with a level of detection of 50 ng of genomic DNA or 10^8 CFU/mL.[136]

If a rapid diagnosis is needed, we suggest choose any of the latest molecular techniques. It is helpful to learn from the following terrific example when a massive surveillance is going to start since the molecular diagnostic tests against comparator methods (bacterial culture, ELISA, and PCR) have already been evaluated in five laboratories worldwide for detecting 15 enteropathogens from 867 diarrheal and 619 non-diarrheal stool specimens.[137] There are other variations of techniques available, most of them are the different versions of PCR For instance, quantitative real-time PCR (qRT-PCR) assays have been used to detect *eltA*, *sta1*, and *sta2* in clinical stool samples with a sensitivity of 89%, significantly more sensitive than that of DNA hybridization assay.[138] In another study, the sensitivity of each of the four qRT-PCR assays was 100% and specificity was ETEC-STIa (92.4%), ETEC-STIb (92.6%), and ETEC-LT (79.6%).[139] The mosaic nature of the genome of some hybrid diarrheagenic strains[46] would make the pathotype categorization difficult.

5.8 TREATMENT AND PREVENTION

There is an empirical management approach for children suffering an acute watery or bloody diarrhea in a resource-limited region,[140] which should be suitable and helpful for managing ETEC diarrhea and infection in general. Prevention measures would be effective if the infrastructure like safe water supply for drinking and irrigation, and the general hygiene is in place and available to residents, a topic beyond the scope of this chapter. For individuals, food safety is the number one measure one should take.

5.8.1 TREATMENT

There are no specific treatments for ETEC-caused diarrhea, and the supportive therapy utilized most is rehydration. For travelers, one should consult doctor before the trip and bring Rifaximin (for the treatment of common afebrile, non-dysenteric diarrhea), a fluoroquinolone, or azithromycin with them in case self-treatment of diarrhea is needed.[141] If the case is more complicated than diarrhea, go to see a local doctor and remember to report to your family physician about the incident when back home to see if there is any need for follow-up examination or treatment. For local people, rehydration is also the first option if the case is without complication.

5.8.2 PREVENTION

For people living locally, certain hygiene standard should be applied for public drinking water supply and in daily food handling and eating/drinking. For tourists, prophylactic measures can be taken and care should be exercised in eating/drinking, sometime sacrificing the desire to taste those delicious but, uncooked/undercooked food including salad, and only drink bottled water.

5.8.3 VACCINATION

A CNN news on March 26, 2016 that the Tribeca Film Festival cancelled a plan to host the world premiere of an anti-vaccine film called "Vaxxed: From Cover-Up to Catastrophe" directed by Andrew Wakefield, an activist and former doctor, partly reflexes the frustration in vaccine development. It is true that not that many vaccines work so perfect, and some with side effects and some without enough or no protection at all to some population.

Although no vaccine to date can provide full protection covering all serotypes, there are a few vaccine candidates in developing and under evaluation, either heat-killed or live attenuated strains, and subunit components. With more than 10 vaccines in either preclinical, early, or advanced stages of clinical development [142,143] including ETVAX (the trade name of a multivalent ETEC vaccine),[144] none of them are currently licensed. The only vaccine available, Dukoral®, is non-ETEC specific, which is an inactivated cholera vaccine and recommended to travelers for prevention of ETEC-associated diarrhea.[145]

It would be optimal if one serotype is prevalent locally or in a place where you are planning to travel to, and there is a vaccine available specifically for that serotype. For researchers, vaccine candidates can be searched based on an in silico analysis of chimeric protein structure and its stability and solubility, with secondary and tertiary structures and mapped T and B cell epitopes.[146] For detail knowledge about vaccine development for ETEC diarrhea, there are reviews recently published.[147,148]

5.8.4 FOOD SAFETY

As a consumer, different strategies can be taken dealing with various eating/drinking scenarios. As a thumb of rule before eating at a restaurant, on the open market or on a cruise ship when travelling to/in an unfamiliar place, dry/wet wash your hands thoroughly first, only drink boiled/bottled water and avoid those uncooked/undercooked items such as sushi, raw meat/fish, and salad. When buying ready-to-eat food to consume or ingredients to cook, do not trust the hygiene too much (we will tell why later), wash the ingredients carefully and fully cook seafood and other foods even at the expense of sacrificing some flavor. Whenever possible, peel the skin of the fruits and then enjoy. If not possible at all (small fruits), wash them really carefully 1st, even with the help of some detergent, and rinse with bottled or boiled water before eating. As a food handler, effective washing of hands, the cooking utensils, and material can dramatically reduce the number of bacteria although not completely eliminating them.[149]

Here is why if possible one should consider to follow the above-mentioned advice while eating. Worldwide it is estimated that through foods 233 million ETEC cases caused 73,041 deaths in 2010 alone.[58] Even ready-to-eat cooked vegetable salads and fresh carrot juice can be contaminated,[150,151] probably during the making process.

Although Vessel Sanitation Program (VSP) was introduced in 1975, from time to time outbreaks from ETEC or other pathogens still occur on board.[152] There have been many ETEC outbreaks due to drinking unbottled water and consuming beverages with ice cubes made from then-unknown contaminated water on ships,[152] eating foods uncooked/undercooked or/and contaminated, including, but not limited to (lettuce) salad,[27,28] sushi,[29] tuna paste,[30] vegetables like kimchi,[19,21] carrot,[31] chives,[20] and parsley,[32]as mentioned earlier.

Cooking ingredients like leafy vegetable and other fresh produce irrigated with water can easily become contaminated because of problematic water source as exemplified by the astonishing report that 100% of surface water, 83% of soil and 67% of vegetable samples were contaminated with at least one of the enteropathogens (ETEC among the most possible) in La Paz River region, Bolivia.[153] A few examples of ETEC-contaminated fresh produces, either domestic or imported, have been reported, including cantaloupe, celery, cilantro, green onion, lettuce, hot peppers, parsley, spinach, sprouts, and tomatoes examined in United States,[154] mint leaves, coriander, methi leaves, and tomato in Mexico,[155–157] raw chicken, pork, and beef in southern Thailand[158] and chilled whole shrimp in Brazil.[159]

5.8.5 CHANGE OF LIFESTYLE

There are several aspects in daily lifestyle one can try to modify/change in order to increase your general resistance to intestinal infection, not necessarily specific for ETEC. Since intestine colonization is the prerequisite of ETEC infection, one can choose some probiotic products that can increase the intestinal colonization resistance by addition of normal flora beneficial strains. For example, some specific strains of *bifidobacterium* and *lactobacillus* can be protective. *Lactobacillus sobrius* (DSM 16698^T) can inhibit IPEC-1 epithelial cells to internalize ETEC, reduce adhesion of ETEC to IPEC-1, and prevent ETEC to cause cell membrane damage by interfering with cytokine production and signal transduction whereas its closely phylogenetic strain *Lactobacillus amylovorus* DSM 20531^T offers no protection.[160]

Change the everyday cuisine ingredients. The transgenic rice expressing LTB (the b subunit of LT) should be given a try if commercially available.[161] Tofu is probably the most popular soya bean products consumed around the world, and extracts of tempe, a fungal fermented soya food is popular in some areas in Asia. Studies show that tempe extracts can reduce ETEC adhesion to Caco-2 cells by 50%,[162] and malnourished children taking meals containing tempe have less episodes and shorter duration of diarrhea than the control kids.[163] There is a very informative review describing plant products with antibacterial and antidiarrheal activities.[164]

To maintain a certain level of some metal ions might be essential to combat infection. For example, supplementation of zinc in food to children with ETEC-caused diarrhea enhances their innate immunity by changing T cell phenotypes, elevating complement C3 and increasing phagocytic activity.[165]

There are actually more stuff one can test (and taste) and the point here exemplified by the few examples listed above is if a change in eating style is not going to harm and may be potentially beneficial to one's health, then why not give it a try? The take home message is take good care of the intestine and it will take care of you.

5.9 FUTURE PERSPECTIVES

Until we have a better vaccine, one has to choose to use what are available. Researchers may develop a panel of vaccines that are region specific, that is, it incorporates epidemiological information into the design and composition of vaccine, although administratively and technically complicated, for tourists and local residents to choose.

Future studies should be done in hospitals discriminating all known enteropathogens and finding out the culprit that is the causal of diarrhea in cases of mixed infections. Although a daunting task, authorities should try to find a way to monitor pathogens in fresh produce and meat in real time in order to take immediate and appropriate action and warn consumers with effective channels. Clinicians will try every effort in treating diarrhea of any kind and authorities of all hierarchical levels should think about changing the current policy of passively receiving patients into actively looking for patients, especially in case of large outbreaks.

KEYWORDS

- **enterotoxin**
- **diarrhea**
- **shiga toxin**
- **pathogens**
- **infections**

REFERENCES

1. Lanata CF., et al. *PLoS One* **2013,** *8 (9)*, e72788.
2. Madhavan, T. P.; Sakellaris, H. *Adv. Appl. Microbiol.* **2015,** *90,* 155.
3. Weiglmeier, P. R.; Rösch, P.; Berkner, H. *Toxins (Basel).* **2010,** *2* (9), 2213.
4. Mudrak, B.; Kuehn, M. J. *Toxins (Basel).* **2010,** *2* (6), 1445.
5. Turner, S. M.; Scott-Tucker, A.; Cooper, L. M.; Henderson, I. R. *FEMS Microbiol. Lett.* **2006,** *263* (1), 10.
6. Qadri, F.; Svennerholm, A. M.; Faruque, A. S.; Sack, R. B. *Clin. Microbiol. Rev.* **2005,** *18* (3), 465.

7. Fleckenstein, J. M.; Hardwidge, P. R.; Munson, G. P.; Rasko, D. A.; Sommerfelt, H.; Steinsland, H. *Microbes Infect.* **2010,** *12* (2), 89.
8. Frank, D. N.; St Amand, A. L.; Feldman, R. A.; Boedeker, E. C.; Harpaz, N.; Pace, N. R. *Proc. Natl. Acad. Sci. USA* **2007,** *104* (34), 13780.
9. Gonzales-Siles, L.; Sjöling, Å. *Environ. Microbiol.* **2016,** *18* (3), 741.
10. Lothigius, A.; Sjöling, A.; Svennerholm, A. M.; Bölin, I. *J. Appl. Microbiol.* **2010,** *108* (4), 1441.
11. Porter, C. K.; Riddle, M. S.; Tribble, D. R.; Louis Bougeois, A.; McKenzie, R.; Isidean, S. D., et al. *Vaccine* **2011,** *29* (35), 5869.
12. McKenzie, R.; Porter, C. K.; Cantrell, J. A.; Denearing, B.; O'Dowd, A.; Grahek, S. L., et al. *J. Infect. Dis.* **2011,** *204* (1), 60.
13. Li, L. L.; Liu, N.; Humphries, E. M.; Yu, J. M.; Li, S.; Lindsay, B. R.; Stine, O. C.; Duan, Z. J. *Clin. Microbiol. Infect.* **2016,** *22* (4), 381.
14. Porter, C. K., et al. *PLoS One* **2016,** *11* (3), e0149358.
15. Skrede, S.; Steinsland, H.; Sommerfelt, H.; Aase, A.; Brandtzaeg, P.; Langeland, N., et al. *BMC Infect. Dis.* **2014,** *14,* 482.
16. Harro, C.; Chakraborty, S.; Feller, A.; DeNearing, B.; Cage, A.; Ram, M.; Lundgren, A., et al. *Clin. Vaccine Immunol.* **2011,** *18* (10), 1719–1727.
17. Coster, T. S., et al. *Infect. Immun.* **2007,** *75* (1), 252–259.
18. Pop, M., et al. *BMC Genomics* **2016,** *17,* 440.
19. Shin, J.; Yoon, K. B.; Jeon, D. Y.; Oh, S. S.; Oh, K. H.; Chung, G. T.; Kim, S. W.; Cho, S. H. *Foodborne Pathog. Dis.* **2016,** *5* (1), 1397.
20. MacDonald, E., et al. *Epidemiol. Infect.* **2015,** *143* (3), 486.
21. Cho, S. H., et al. *Epidemiol. Infect.* **2014,** *142* (3), 616.
22. Beatty, M. E., et al. *Clin. Infect. Dis.* **2006,** *42* (3), 329.
23. O'Mahony, M.; Noah, N. D.; Evans, B.; Harper, D.; Rowe, B.; Lowes, J. A.; Pearson, A.; Goode, B. *J. Hyg. (Lond).* **1986,** *97* (2), 229.
24. Rosenberg, M. L., et al. *Ann. Intern. Med.* **1977,** *86* (6), 714.
25. Kimura, T. I.; Akiba, Y.; Tsuruta, M.; Akimoto, T.; Mitsui, Y.; Ogasawara, Y.; Ikegami, H. *Jpn. J. Infect. Dis.* **2006,** *59* (6), 410.
26. Dong, Y. L.; Yang, Q. Z.; Chen, S. J. *J. Hosp. Infect.* **2010,** *74* (4), 405.
27. Pakalniskiene. J., et al. *Epidemiol. Infect.* **2009,** *137* (3), 396.
28. Ethelberg, S., et al. *Euro. Surveill.* **2010,** *15* (6), 19484.
29. Jain, S., et al. *Clin. Infect. Dis.* **2008,** *47* (1), 1.
30. Mitsuda, T.; Muto, T.; Yamada, M.; Kobayashi, N.; Toba, M.; Aihara, Y.; Ito, A.; Yokota, S. *J. Clin. Microbiol.* **1998,** *36* (3), 652.
31. No author listed. *JAMA* **1994,** *271* (9), 652.
32. Naimi, T. S., et al. *J. Food Prot.* **2003,** *66* (4), 535.
33. Yoder, J. S.; Cesario, S.; Plotkin, V.; Ma, X.; Kelly-Shannon, K.; Dworkin, M. S. *Clin. Infect. Dis.* **2006,** *42* (11), 1513.
34. Rowe, B.; Gross, R. J.; Scotland, S. M.; Wright, A. E.; Shillom, G. N.; Hunter, N. J. *J. Clin. Pathol.* **1978,** *31* (3), 217.
35. Harada, T., et al. Jpn. *J. Infect. Dis.* **2013,** *66* (6), 530.
36. Devasia, R. A., et al. *Am. J. Med.* **2006,** *119* (2), 168.e7–168.e10.
37. Taneja, N.; Das, A.; Raman Rao, D. S.; Jain, N.; Singh, M.; Sharma, M. *J. Hosp. Infect.* **2003,** *53* (3), 193.

38. Beatty, M. E.; Bopp, C. A.; Wells, J. G.; Greene, K. D.; Puhr, N. D.; Mintz, E. D. *Emerg. Infect. Dis.* **2004,** *10* (3), 518.
39. Chakraborty, S., et al. *J. Clin. Microbiol.* **2001,** *39* (9), 3241.
40. Liu, J., et al. *Lancet* **2016,** *388* (10051), 1291.
41. Sahl, J. W., et al. *MBio.* **2015,** *6* (3), e00501.
42. Wang, X., et al. *PLoS One* **2015,** *10* (11), e0142136.
43. Lindsay, B. R., et al. *FEMS Microbiol. Lett.* **2014,** *352* (1), 25.
44. Lääveri, T.; Antikainen, J.; Pakkanen, S. H.; Kirveskari, J.; Kantele, A. *Clin. Microbiol. Infect.* **2016,** *22* (6), 535.
45. Wyrsch, E., et al. *BMC Genomics*. **2015,** *16,* 165.
46. Dutta, S.; Pazhani, G. P.; Nataro, J. P.; Ramamurthy, T. *Int. J. Med. Microbiol.* **2015,** *305* (1), 47.
47. Martikainen, O.; Kagambèga, A.; Bonkoungou, I. J.; Barro, N.; Siitonen, A.; Haukka, K. *Foodborne Pathog. Dis.* **2012,** *9* (11), 1015.
48. Nyholm, O.; Heinikainen, S.; Pelkonen, S.; Hallanvuo, S.; Haukka, K.; Siitonen, A. *Zoonoses Public Health.* **2015,** *62* (7), 518.
49. Leonard, S. R.; Mammel, M. K.; Rasko, D. A.; Lacher, D. W. *Appl. Environ. Microbiol.* **2016,** *82* (14), 4309.
50. Nyholm, O., et al. *PLoS One* **2015,** *10* (8), e0135936.
51. Medus, C., et al. *Open Forum Infect. Dis.* **2016,** *3* (1), ofw003.
52. Bhavnani, D., et al. *Am. J. Trop. Med. Hyg.* **2016,** *94* (2), 276.
53. Spina, A., et al. *Clin. Microbiol. Infect.* **2015,** *21* (8), 719.
54. Khare, R., et al. *J. Clin. Microbiol.* **2014,** *52* (10), 3667.
55. Tobias, J., et al. *BMC Infect. Dis.* **2015,** *15,* 79. doi: 10.1186/s12879-015-0804-4
56. Zboromyrska, Y., et al. *Clin. Microbiol. Infect.* **2014,** *20* (10), O753.
57. Lamberti, L. M.; Bourgeois, A. L.; Fischer Walker, C. L.; Black, R. E.; Sack, D. *PLoS. Negl. Trop. Dis.* **2014,** *8* (2), e2705.
58. Pires, S. M., et al. *PLoS One* **2015,** *10* (12), e0142927.
59. Lozano, R.; Naghavi, M.; Foreman, K.; Lim, S.; Shibuya, K.; Aboyans, V., et al. *Lancet* **2013,** *380,* 2095.
60. Johnson, T. J.; Nolan, L. K. *Microbiol. Mol. Biol. Rev.* **2009,** *73* (4), 750.
61. Tamhankar, A. J., et al. *Genome Announc.* **2015,** *3* (2), e00225–15.
62. Crossman, L. C., et al. *J. Bacteriol.* **2010,** *192* (21), 5822.
63. Ochi, S.; Shimizu, T.; Ohtani, K.; Ichinose, Y.; Arimitsu, H.; Tsukamoto, K.; Kato, M.; Tsuji, T. *DNA Res.* **2009,** *16* (5), 299.
64. Sahl, J. W.; Steinsland, H.; Redman, J. C.; Angiuoli, S. V.; Nataro, J. P.; Sommerfelt, H.; Rasko, D. A. *Infect. Immun.* **2011,** *79* (2), 950.
65. von Mentzer, A., et al. *Nat. Genet.* **2014,** *46* (12), 1321.
66. Turner, S. M., et al. *J. Clin. Microbiol.* **2006,** *44* (12), 4528.
67. Pettersen, V. K.; Steinsland, H.; Wiker, H. G. *Proteomics.* **2015,** *15* (22), 3826.
68. Begum, Y. A.; Baby, N. I.; Faruque, A. S.; Jahan, N.; Cravioto, A.; Svennerholm, A. M.; Qadri, F. *PLoS. Negl. Trop. Dis.* **2014,** *8* (7), e3031.
69. Andersson, M.; Björnham, O.; Svantesson, M.; Badahdah, A.; Uhlin, B. E.; Bullitt, E. *J. Mol. Biol.* **2012,** *415* (5), 918.
70. Roy, K.; Hilliard, G. M.; Hamilton, D. J.; Luo, J.; Ostmann, M. M.; Fleckenstein, J. M. *Nature.* **2009,** *457* (7229), 594.

71. Sheikh, A.; Luo, Q.; Roy, K.; Shabaan, S.; Kumar, P.; Qadri, F.; Fleckenstein, J. M. *Infect. Immun.* **2014,** *82* (9), 3657.
72. Allen, K. P.; Randolph, M. M.; Fleckenstein, J. M. *Infect. Immun.* **2006,** *74* (2), 869.
73. Ghosal, A., et al. *Infect. Immun.* **2009,** *77* (5), 2125.
74. Jansson, L.; Tobias, J.; Jarefjäll, C.; Lebens, M.; Svennerholm, A. M.; Teneberg, S. *PLoS One* **2009,** *4* (2), e4487.
75. Sabui, S., et al. *J. Med. Microbiol.* **2010,** *59* (Pt 7), 770.
76. Debnath, A.; Wajima, T.; Sabui, S.; Hamabata, T.; Ramamurthy, T.; Chatterjee, N. S. *Microbiology* **2015,** *161* (Pt 4), 865.
77. Joffré, E.; Sjöling, Å. *Gut Microbes.* **2016,** *7* (1), 75.
78. Lasaro, M. A., et al. *FEMS Immunol. Med. Microbiol.* **2006,** *48* (1), 123.
79. Kunkel, S. L.; Robertson, D. C. *Infect. Immun.* **1979,** *23* (3), 652.
80. Gonzales, L., et al. *PLoS One* **2013,** *8* (9), e74069.
81. Sjöling, A.; Wiklund, G.; Savarino, S. J.; Cohen, D. I.; Svennerholm, A. M. *J. Clin. Microbiol.* **2007,** *45* (10), 3295.
82. Haycocks, J. R.; Sharma, P.; Stringer, A. M.; Wade, J. T.; Grainger, D. C. *PLoS Pathog.* **2015,** *11* (1), e1004605.
83. Trachman, J. D.; Maas, W. K. *J. Bacteriol.* **1998,** *180* (14), 3715.
84. Tauschek, M.; Gorrell, R. J.; Strugnell, R. A.; Robins-Browne, R. M. *Proc. Natl. Acad. Sci. USA.* **2002,** *99* (10), 7066.
85. Fleckenstein, J. M.; Lindler, L. E.; Elsinghorst, E. A.; Dale, J. B. *Infect. Immun.* **2000,** *68* (5), 2766.
86. Horstman, A. L.; Kuehn, M. J. *J. Biol. Chem.* **2002,** *277* (36), 32538.
87. Takashi, K.; Fujita, I.; Kobari, K. *Jpn. J. Pharmacol.* **1989,** *50* (4), 495.
88. Spangler, B. D. *Microbiol. Rev.* **1992,** *56* (4), 622.
89. Casey, T. A.; Connell, T. D.; Holmes, R. K.; Whipp, S. C. *Vet. Microbiol.* **2012,** *159*(1–2), 83.
90. Hofstra, H.; Witholt, B. *J. Biol. Chem.* **1985,** *260* (29), 16037.
91. Yu, J.; Webb, H.; Hirst, T. R. *Mol. Microbiol.* **1992,** *6* (14), 1949.
92. Moss, J.; Osborne, J. C., Jr.; Fishman, P. H.; Nakaya, S.; Robertson, D. C. *J. Biol. Chem.* **1981,** *256* (24), 12861.
93. Flinn, J. P.; Pallaghy, P. K.; Lew, M. J.; Murphy, R.; Angus, J. A.; Norton, R. S. *Biochim. Biophys. Acta.* **1999,** *1434* (1), 177.
94. Sears, C. L.; Kaper, J. B. *Microbiol. Rev.* **1996,** *60* (1), 167.
95. Nataro, J. P.; Kaper, J. B. *Clin. Microbiol. Rev.* **1998,** *11* (1), 142.
96. Moseley, S. L.; Samadpour-Motalebi, M.; Falkow, S. *J. Bacteriol.* **1983,** *156* (1), 441.
97. Vaandrager, A. B, *Mol. Cell. Biochem.* **2002,** *230*(1–2), 73.
98. Vaandrager, A. B., et al. *J. Biol. Chem.* **1997,** *272* (7), 4195.
99. Vaandrager, A. B., et al. *Proc. Natl. Acad. Sci. USA* **1998,** *95* (4), 1466.
100. Khare, S., et al. *Arch. Biochem. Biophys.* **1994,** *314* (1), 200.
101. Harville, B. A.; Dreyfus, L. A. *Infect. Immun.* **1995,** *63* (3), 745.
102. Read, L. T., et al. *Infect. Immun.* **2014,** *82* (12), 5308.
103. Turner, A. K., et al. *Infect. Immun.* **2001,** *69* (8), 4969.
104. Fleckenstein, J. M.; Lindler, L. E.; Elsinghorst, E. A.; Dale, J. B. *Infect. Immun.* **2000,** *68* (5), 2766.
105. Horstman, A. L.; Kuehn, M. J. *J. Biol. Chem.* **2000,** *275* (17), 12489.

106. Roy, K.; Hamilton, D.; Allen, K. P., et al. *Infect. Immun.* **2008,** *76*(5), 2106.
107. Roy, K.; Hamilton, D. J.; Fleckenstein, J. M. *Clin. Vaccine Immunol.* **2012,** *19* (10), 1603.
108. Luo, Q., et al. *Infect. Immun.* **2014,** *82* (2), 509.
109. Roy, K., et al. *J. Biol. Chem.* **2011,** *286* (34), 29771.
110. Mammarappallil, J. G.; Elsinghorst, E. A. *Infect. Immun.* **2000,** *68* (12), 6595.
111. Lindenthal, C.; Elsinghorst, E. A. *Infect. Immun.* **2001,** *69* (1), 52.
112. Yamamoto, T.; Echeverria, P. *Infect. Immun.* **1996,** *64* (4), 1441.
113. Haines, S.; Gautheron, S.; Nasser, W.; Renauld-Mongénie, G. *PLoS One* **2015,** *10* (10), e0141469.
114. Sahl, J. W.; Rasko, D. A. *Infect. Immun.* **2012,** *80* (3), 1232.
115. Nicklasson, M., et al. *PLoS One* **2012,** *7* (4), e35827.
116. Wang, X.; Li, S.; Lu, X., et al. *Mol. Med. Rep.* **2016,** *13* (1), 257.
117. Sturbelle, R. T., et al. *Vet. Microbiol.* **2015,** *180*(3–4), 245.
118. Bodero, M. D.; Munson, G. P. *Infect. Immun.* **2009,** *77* (2), 791.
119. Wijemanne, P.; Moxley, R. A. *PLoS One* **2014,** *9* (11), e113230.
120. Karjalainen, T. K., et al. *Microb .Pathog.* **1991,** *11* (5), 317.
121. Haines, S., et al. *J. Bacteriol.* **2015,** *197* (18), 2896.
122. Lu, X.; Fu, E.; Xie, Y.; Jin, F. *Infect. Immun.* **2016,** *84* (10), 2748–57.
123. Kansal, R., et al. *Infect. Immun.* **2013,** *81* (1), 259.
124. Sugarman, B.; Epps, L. R. *J. Med. Microbiol.* **1984,** *18* (3), 393.
125. Levner, M.; Wiener, F. P.; Rubin, B. A. *Infect. Immun.* **1977,** *15* (1), 132.
126. Wang, X.; Gao, X.; Hardwidge, P. R. *Cell. Microbiol.* **2012,** *14* (8), 1231.
127. Wang, X.; Hardwidge, P. R. *Infect. Immun.* **2012,** *80* (12), 4417.
128. Finamore, A., et al. *PLoS One* **2014,** *9* (4), e94891.
129. Ren, W., et al. *Microbes. Infect.* **2014,** *16* (11), 954.
130. Lai. X. H.; Xu, J. G.; Melgar, S.; Uhlin, B. E. *FEMS Microbiol. Lett.* **1999,** *172* (1), 29.
131. Tsuji, T.; Asano, Y.; Handa, T., et al. *Immunobiol.* **2000,** *201* (3–4), 377.
132. Syed, H. C.; Dubreuil, J. D. *Microb. Pathog.* **2012,** *53* (3–4), 147.
133. Tang, Y.; Li, F.; Tan, B., et al. *Vet. Microbiol.* **2014,** *171* (1–2), 160.
134. Zhou, C.; Liu, Z.; Jiang, J.; Yu, Y.; Zhang, Q. *BMC Genomics* **2012,** *13,* 330.
135. Yang, W. E., et al. *J. Infect. Dis.* **2016,** *213* (9), 1495.
136. Wang, Q.; Wang, S.; Beutin, L.; Cao, B.; Feng, L.; Wang, L. *J. Clin. Microbiol.* **2010,** *48* (6), 2066.
137. Liu, J., et al. *Lancet Infect. Dis.* **2014,** *14* (8), 716.
138. Youmans, B. P., et al. *Am. J. Trop. Med. Hyg.* **2014,** *90* (1), 124.
139. Silapong, S., et al. *US Army Med. Dep. J.* **2015,** 51–58.
140. O'Ryan, G. M., et al. *Expert Rev. Anti. Infect. Ther.* **2014,** *12* (5), 621.
141. Sack, R. B. *Clin. Infect. Dis.* **2005,** *41*(Suppl 8), S553.
142. O'Ryan, M., et al. *Hum. Vaccin Immunother.* **2015,** *11* (3), 601.
143. Bourgeois, A. L.; Wierzba, T. F.; Walker, R. I. *Vaccine* **2016,** *34* (26), 2880.
144. Lundgren, A.; Jertborn, M.; Svennerholm, A. M. *Vaccine* **2016,** *34* (27), 3132.
145. Ahmed, T., et al. *Cochrane Database Syst. Rev.* **2013,** *7,* CD009029.
146. Jeshvaghani, F. S., et al. *Protein Pept. Lett.* **2016,** *23* (1), 33.
147. Zhang, W.; Sack, D. A. *Clin. Vaccine Immunol.* **2015,** *22* (9), 983.

148. Fleckenstein, J.; Sheikh, A.; Qadri, F. *Expert Rev. Vaccines* **2014,** *13* (5), 631.
149. Abdelnoor, A. M., et al. *Zentralbl Bakteriol. Mikrobiol. Hyg. B* **1983,** *177* (3–4), 342.
150. Torres-Vitela, M. D., et al. *Lett. Appl. Microbiol.* **2013,** *56* (3), 180.
151. Bautista-De León, H., et al. *J. Lett. Appl. Microbiol.* **2013,** *56* (6), 414.
152. Daniels, N. A., et al. *J. Infect. Dis.* **2000,** *181* (4), 1491.
153. Poma, V.; Mamani, N.; Iñiguez, V. *Springerplus*. **2016,** *5,* 499.
154. Feng, P. C.; Reddy, S. P. *J. Food Prot.* **2014,** *77* (5), 820.
155. Gómez-Aldapa, C. A., et al. *J. Food Prot.* **2013,** *76* (9), 1621.
156. Gómez-Aldapa, C. A., et al. *Food Microbiol.* **2016,** *59,* 97.
157. Singh, G.; Vajpayee, P.; Ram, S.; Shanker, R. *Environ. Sci. Technol.* **2010,** *44* (16), 6475.
158. Phetkhajorn, S., et al. *Southeast Asian J. Trop. Med. Public Health.* **2014,** *45* (6), 1385.
159. Barbosa, L. J., et al. *Lett. Appl. Microbiol.* **2016,** *62* (5), 372.
160. Roselli, M., et al. *J. Nutr.* **2007,** *137* (12), 2709.
161. Soh, H. S., et al. *Springerplus* **2015,** *4,* 148.
162. Roubos-van den Hil, P. J., et al. *J. Appl. Microbiol.* **2009,** *106* (3), 1013.
163. Kalavi, F. N.; Muroki, N. M.; Omwega, A. M.; Mwadime, R. K. *East Afr. Med. J.* **1996,** *73* (7), 427.
164. Dubreuil, J. D. *Toxins (Basel)* **2013,** *5* (11), 2009.
165. Sheikh, A., et al. *J. Nutr.* **2010,** *140* (5), 1049.

CHAPTER 6

LISTERIA MONOCYTOGENES AS FOODBORNE PATHOGEN: GENETIC APPROACHES, IDENTIFICATION, AND DETECTION METHODS

HOSSAM ABDELHAMED, SEONGWON NHO, ATTILA KARSI, and MARK L. LAWRENCE*

Department of Basic Sciences, College of Veterinary Medicine, Mississippi State University, Starkville 39762, MS, USA

**Corresponding author. E-mail: Lawrence@cvm.msstate.edu*

CONTENTS

Abstract 152
6.1 Introduction 152
6.2 Biological Characteristics 153
6.3 Disease 154
6.4 Epidemiology 154
6.5 Genomics 156
6.6 Pathogenesis 165
6.7 Identification and Detection 166
6.8 Conclusion and Future Research 170
Keywords 170
References 171

ABSTRACT

Listeria monocytogenes is an intracellular pathogen transmitted to humans and animals through the consumption of contaminated food. It causes listeriosis and affects mostly at-risk people causing high hospitalization rates and death. To develop improved control measures for listeriosis, it is important to understand pathogenic mechanisms of *L. monocytogenes*. Manipulating chromosomal genes is a fundamental biological tool for the analysis of gene function in bacterial species. In this chapter, we summarize listeriosis, characteristic features of *Listeria* species, and sources and routes of transmission of *Listeria* species. Also, we discuss the approaches and tools available for gene inactivation in *L. monocytogenes* by allelic exchange and homologous recombination. We also detail the construction of a novel suicide delivery system (pHoss1). Finally, we review available laboratory identification and detection methods for *Listeria* species.

6.1 INTRODUCTION

Illnesses caused by the consumption of contaminated food have a global public health impact. Microbial pathogens in contaminated food have been estimated to cause 6.5–33 million cases of human illness and up to 9000 deaths in the United States each year. More than 40 different foodborne pathogens, including fungi, viruses, parasites, and bacteria are known to cause human foodborne illnesses. Bean et al.[108] found that over 90% of confirmed foodborne human illness cases and deaths reported to the Centers for Disease Control and Prevention (CDC) have bacterial origin. The foodborne bacterial cases were caused mostly by the following five pathogens: *Campylobacter* species, *Salmonella enterica, Yersinia enterocolitica,* Shiga toxin-*Escherichia coli* (STEC), and *Listeria monocytogenes*. The costs of these illnesses are estimated to be $9.3–$12.9 billion annually.[1]

L. monocytogenes causes listeriosis, a serious infection with high hospitalization and mortality rates in at-risk populations. *L. monocytogenes* differs from most foodborne pathogens in that it is ubiquitous in the environment, resistant to wide range of environmental conditions including extremely low pH and high salt concentrations, and has psychotropic and mesophilic capabilities. This chapter describes *Listeria* species as a foodborne pathogen. Included is a description of approaches and tools available for targeted mutagenesis by allelic exchange. We also describe construction of a new suicide delivery system (pHoss1). Finally, we review available laboratory identification and detection methods for the etiological diagnosis of *Listeria* species.

6.2 BIOLOGICAL CHARACTERISTICS

Listeria species are short rods measuring 0.4–0.5 μm by 0.5–2.0 μm with rounded ends.[2] They are Gram-positive, nonspore-forming, and facultative anaerobic bacteria. *L. monocytogenes* is motile due to peritrichous flagella, which may not be expressed as the bacteria enter eukaryotic cells. *L. monocytogenes* can grow intracellularly within monocytes and neutrophils.[3]

The genus *Listeria* was classified in the family of *Corynebacteriaceae*, but based on 16S rRNA sequencing; it has been placed in the newly created family *Listeriaceae* since 2001. The genus *Listeria* has ten named species: *Listeria fleischmannii, Listeria grayi, Listeria. innocua, Listeria ivanovii, Listeria marthii, L. monocytogenes, Listeria rocourtiae, Listeria seeligeri, Listeria weihenstephanensis,* and *Listeria Welshimeri.*[4] *L. monocytogenes* is the only member of the genus *Listeria* that causes disease in human and animals. *L. ivanovii* causes disease in animals only, mainly sheep. Encephalitis is the most common manifestation of the disease in ruminant animals.[109] On the basis of somatic (O) and flagellar (H) antigens, *L. monocytogenes* is composed of 13 serotypes including 1/2a, 1/2b, 1/2c, 3a, 3b, 3c, 4a, 4ab, 4b, 4c, 4d, 4e, and 7. However, only three serotypes (1/2a, 1/2b, and 4b) are reported to be responsible for the vast majority of human foodborne infections.[5]

L. monocytogenes can adapt to survive and grow in adverse environmental conditions such as low temperatures, high salt concentrations, and low pH. *L. monocytogenes,* unlike many other pathogens, is capable of surviving in temperatures from −7 to 45°C.[6] The optimum growth temperature for *L. monocytogenes* is around 30 and 37°C, but it is capable of growth at temperatures as low as 0°C.[7] As a result, it can multiply in food under refrigerated temperatures. *Listeria* can adapt and survive in high concentrations of salt, which makes it difficult to control in foods. In *L. monocytogenes,* 12 proteins induced by salt stress were identified by comparative two-dimensional polyacrylamide gel electrophoresis.[8] *L. monocytogenes* can form biofilms on solid surfaces, including food-processing facilities.

In this state, *L. monocytogenes* is more resistant to disinfectants and sanitizing agents.

6.3 DISEASE

L. monocytogenes is the causative agent of listeriosis, an important public health problem worldwide with a significant risk of mortality and morbidity. The disease may occur in non-invasive and invasive forms.[9] Non-invasive listeriosis (febrile gastroenteritis) is the milder form of the disease and occurs in healthy individuals. The symptoms of this illness include diarrhea, fever, headache, muscle aches, vomiting, and myalgia.[5] Symptoms occur after a short incubation period that ranges from a few hours to 2 or 3 days. Outbreaks of this disease require ingestion of a large number of viable *L. monocytogenes.* Invasive listeriosis is the severe form of the disease and affects high-risk groups including pregnant women, immunocompromised adults, and senior citizens. This disease is characterized by an invasive systemic infection leading to sepsis, meningitis, and meningoencephalitis with a high mortality rate (25–30%).[10] Also, fetal infections cause spontaneous abortions, stillbirth, premature labor, and neonatal disease.[11] The incubation period for invasive illness is not well established; evidence from a few cases suggests incubation can range from 3 days to 3 months, with a mean of 31 days.[12]

It is estimated that 5% of clinically healthy individuals carry *L. monocytogenes* in their intestinal tract without signs of disease.[13] The Economic Research Services (ERS) estimated that the cost of acute illness from foodborne *L. monocytogenes* is $2.3 billion each year in the United States.[14] The CDC estimates 260 deaths annually due to foodborne *L. monocytogenes* in the United States. The severe form of listeriosis has a case-fatality rate of up to 30% overall. When listerial meningitis occurs, the case-fatality rate may be as high as 70%; when septicemia occurs, it may be 50%; and in perinatal/neonatal infections it can be more than 80%. Outbreaks of listeriosis and contaminated product recalls continue to occur.[15]

6.4 EPIDEMIOLOGY

L. monocytogenes is widespread in nature and an important cause of zoonoses. It has been isolated from soil, water, silage, decaying vegetation, and other environmental sources.[16] Animals can carry *L. monocytogenes* asymptomatically and can contaminate foods. *L. monocytogenes* is a part of the fecal flora of a wide variety of animal species and some species of fish and shellfish. Reservoirs animal include cattle, sheep, and avian species. Humans also act as reservoirs, and asymptomatic fecal carriage has been reported. It is estimated that 5% of healthy individuals carry *L. monocytogenes* in their intestinal tract asymptomatically.[13]

L. monocytogenes is found commonly in food, and it can contaminate a wide variety of raw and processed foods. However, the processing environment was considered to be the primary source of contamination of food products. Foods implicated in outbreaks with *L. monocytogenes* include milk (especially unpasteurized milk), other dairy products such as cheeses (particularly soft cheeses) and ice cream, various meat and meat products (smoked seafood, beef, pork, and fermented sausages), raw vegetables such as radishes and cabbage, and seafood products.[17]

The first case of human infection with *L. monocytogenes* was reported in 1929.[3] However, the incidence of listeriosis remained rare until the 1980s when a series of outbreaks were reported in the United States and Canada.[18,19] Since this outbreak, many incidents of *L. monocytogenes* foodborne infection have been reported worldwide. The first recognized outbreaks of listeriosis in Europe and the United States were due to consumption of contaminated milk products.[20] Many incidences of listeriosis have occurred during the last decade in Europe and the United States.[21] From 1991 to 2002, a total of 19 outbreaks of invasive *Listeria* infection were reported in nine different countries in Europe, with a total of 526 outbreak-related cases.[22] In 1997, a large outbreak occurred in northern Italy due to the consumption of contaminated corn salad. This outbreak resulted in 1566 cases of febrile gastroenteritis illness, and a total of 292 persons were hospitalized.[23] According to Health Protection Surveillance Centre (HPSC), an increase in the incidence of listeriosis in many European countries including England and Wales, Denmark, Belgium, Germany, the Netherlands, Switzerland, and Finland was observed during 2000–2006.

In the United States, listeriosis incidence has declined in recent years. The estimated incidence decreased significantly by 24% from 4.1 cases per million people in 1996 to 3.1 cases per million people in 2003. Listeriosis incidence has declined by 42% with rates reaching 0.27 per 100,000 in 2007 compared with the baseline period (1996–1998). The average annual incidence of listeriosis was 0.26 cases per 100,000 individuals in

2013.[24] However, despite this decrease, foodborne outbreaks remain a serious problem. The outbreak associated with consumption of cantaloupe in 2011 was the largest listeriosis outbreak in United States history.[25]

Australia's national notifiable disease surveillance system reported average annual listeriosis incidence rates ranging from 35 to 73 cases during the period 1991–2007. In Singapore, the annual number of patient required notifications to the Ministry of Health ranged from one to nine cases from 2001 to 2007.[26]

6.5 GENOMICS

Based on somatic and flagellar antigens, *L. monocytogenes* is classified into 13 serotypes[27] of which serotypes 1/2a, 1/2b, and 4b represent the majority of disease-causing strains.[28] Phylogenetic analyses have shown that *L. monocytogenes* serotypes are divided into lineages I, II, and III.[29,30] Lineage I is composed of strains from serotypes 4b, 3b, and 1/2b,[31] lineage II consists of serotypes 1/2a, 1/2c, and 3a,[32] and lineage III includes serotypes 4a and 4c.[33] A subgroup of lineage III, lineage IIIB, was later separated into an independent lineage (lineage IV).[34] These serologic and genetic subtypes are clinically important; over 98% of sporadic cases and epidemic outbreaks caused by *L. monocytogenes* are due serotypes in lineages I and II (4b, 1/2a, 1/2b, and 1/2c). Lineage III strains are not as common in foods and the environment, and they are not related to disease outbreaks in humans.[35] Lineage IV strains are rarely isolated; however, they include strains isolated from humans, animals, and food.

Genome sequences of *L. monocytogenes* isolates (pathogenic and nonpathogenic) led to discoveries in virulence factor regulation, listerial physiology, adaptation to host cytosol, and disease pathogenesis. Currently, NCBI's database contains 49 complete and 236 draft or incomplete genomes (either contigs or scaffolds), and this number continues to increase.36 The complete *L. monocytogenes* genome sequences include strain EGD-e (serotype 1/2a);37 outbreak *strains* F2365 (serotype 4b), F6854 (serotype 1/2a), and H7858 (serotype 4b);38 and serotype 4a strain HCC23.39 Other *Listeria* strains include *L. innocua* CLIP 11262 (serotype 6a),37 *L. welshimeri* SLCC 5334 (serotype 6b), *L. seeligeri* SLCC 3954 (serotype 1/2b), and *L. ivanovii* PAM 55 (serotype 5) (http://www.genomesonline.org).

Comparative genome analysis between *L. monocytogenes* strain EGD-e and nonpathogenic *L. innocua* strain CLIP1182 revealed potential genetic differences responsible for the virulence of this pathogen.[37] A more comprehensive comparison using lineage I strain F2365, lineage II strain EGD-e, lineage III strain HCC23, and *L. innocua* CLIP1182 revealed 58 proteins unique to F2365 and EGD-e compared to nonpathogenic strains HCC23 and CLIP1182.[39] Genomic comparison between strains F2365 (serotype 4b) and EGD-e (serotype 1/2a) revealed a significant number of single nucleotide polymorphisms and some gene additions/deletions.[38] Comparative genomics analysis of *Listeria* species revealed species-specific adaptations.[40] Application of the listerial "pan-genome" enabled identification of lineage-specific *L. monocytogenes* genes, particularly in carbohydrate utilization and stress resistance.[41]

6.5.1 GENETIC MANIPULATION

Since 1986, several techniques have become available for genetic studies on *L. monocytogenes*. Transposons including Tn1545 and Tn916 allow construction of random mutations in the *Listeria* chromosome.[42] Plasmid vectors enabling allelic exchange and site-directed mutagenesis were also developed. In the following section, we will briefly discuss some approaches and tools available for gene inactivation in *L. monocytogenes* by allelic exchange. In particular, we will describe the construction of a new suicide plasmid, pHoss1.

Isogenic mutant strains are important to provide insight into gene functions. Targeted gene disruption allows the study of specific pathways and has been a very useful technique in virulence gene identification and development of attenuated vaccines. Mutation of specific amino acid residues by site-directed mutagenesis enables more detailed analysis of protein function.

Allelic exchange by homologous recombination is a procedure used for the replacement of a wild-type chromosomal allele with a cloned allele that has been manipulated in vitro.[43] This procedure includes in-frame deletions and point mutations. When no antibiotic resistance marker is inserted, the mutation is defined as markerless genetic exchange.[44] One advantage of allelic exchange is preventing potential polar effects of down-stream genes by the placement of a targeted gene in its natural location in the chromosome, which is optimal for analysis of the gene mutation's effects.[45] If an in-frame deletion is constructed, the allelic replacement does not affect the expression of downstream genes in operons, thus avoiding polar effects. Markerless exchange also avoids introduction of antibiotic resistance genes, which are undesirable in vaccine strains. Also, markerless exchange enables multiple mutations that can be serially introduced into the same strain.

Allelic exchange is achieved through a two-step integration-segregation strategy. In this process, the homologous regions upstream and downstream of the target gene are cloned into a suicide plasmid. In the first integration step, the plasmid harboring an in-frame deletion of the mutated gene is integrated into the chromosome by selection for a plasmid-encoded antibiotic resistance gene. In the second segregation step, the gene duplication in the merodiploid strain is resolved by a second recombination, which results in loss of the integrated plasmid in the chromosome. The resulting recombinant strain carries either the wild-type allele or the mutated allele depending on the site of recombination.[45]

To accelerate the identification of desired knockout candidates after allelic replacement, counter-selectable markers incorporated into the plasmid backbone have been employed.[43] This strategy enables positive selection, rather than laborious screening for loss of a phenotype. Isolating mutants using two-step allele exchange can be labor intensive without a counter-selection marker due to the low frequency of the second recombination event. However, identification of a suitable counter-selection marker is complex and usually species-specific.[46]

Under appropriate growth conditions, a counter-selectable marker promotes the death of the bacteria carrying that gene. Some commonly used counter-selectable markers encode gene products that function by (1) inhibition of bacterial growth in the presence of sucrose (mediated by the *sacB* gene product),[47] (2) lipophilic chelators such as fusaric acid mediated by gene products conferring tetracycline resistance,[48] and (3) inhibition of growth by purine or pyrimidine analogs (mediated by phosphoribosyltransferases of the purine and pyrimidine base salvage pathways).[44] These counter-selectable markers have been used to construct deletion mutants in *Mycobacterium tuberculosis*, *Helicobacter pylori*, *Bordetella pertussis*, and several other species. In all of these examples, the counter-selectable marker is carried in the plasmid; thus, when recombinant bacteria are placed in medium containing the appropriate counter-selective agent, positive selection occurs for loss of the integrated plasmid.

6.5.2 APPROACHES FOR L. MONOCYTOGENES MUTATION

Many vectors are available to use for in-frame deletion in *L. monocytogenes* through allelic replacement. These vectors carry a thermo-sensitive Gram-positive origin of replication; most use a thermo-sensitive mutation of the replication origin from plasmid pF194. The vectors most often used for in-frame deletion of *L. monocytogenes* are provided in Table 6.1. For each of these plasmids, recombinant plasmids carrying a mutated gene are introduced into *Listeria*, and bacteria are incubated on medium with the appropriate antibiotic selection at a permissive temperature for plasmid replication (30°C). *L. monocytogenes* with the chromosomal integration of the plasmid are then selected by a second incubation step at a nonpermissive temperature for plasmid replication (40–42°C). This leads to the chromosomal integration of the recombinant plasmid by homologous recombination between the plasmid-borne mutated gene and the chromosomal wild-type gene.

TABLE 6.1 Thermosensitive Suicide Vectors Used for Mutagenesis by Allelic Exchange in *L. monocytogenes*.

Plasmid	Size	Description	Reference
pAUL-A	9.2	Allelic exchange, Em^r	[49]
pCON1	7.6	Allelic exchange, Cm^r	[50]
pKSV7	6.9	Allelic exchange, Cm^r	[51]
pMAD	9.7	Allelic exchange, Em^r	[52]
pLSV2	6.4	Allelic exchange, Em^r	[53]

Recombinant *L. monocytogenes* is then cultivated in the absence of antibiotic selection. After a variable number of replication cycles, spontaneous excision of the integrated plasmid occurs by a second recombination event in the merodiploid strain. Screening for the second recombination event is labor intensive and requires identification of colonies that have lost plasmid-mediated antibiotic resistance. The second recombination results in antibiotic sensitive colonies that are either deletion mutants or wild-type revertants.

Thermosensitive suicide vectors pAUL-A and pLSV2[49,53] have high-efficiency integration into the listerial chromosome (the first recombination event). However, these plasmids do not have a selection marker for the second allelic exchange. Although deletion mutants have been constructed with these plasmids, screening for plasmid excision (the second recombination event) is very labor intensive and requires polymerase chain reaction (PCR) screening of a large number of colonies to identify a single deletion mutant among several hundred wild-type revertants.[46]

pMAD was developed for generating deletion mutations by allelic replacement in several types of Gram-positive bacteria, including *Staphylococcus aureus, L. monocytogenes*, and *Bacillus cereus*[52] In addition to a temperature sensitive origin of replication and an erythromycin resistance selection marker (similar to pAUL-A and pLSV2), pMAD also carries a *lacZ* gene (encoding β-galactosidase) for blue-white screening. Color screening assists in identifying antibiotic sensitive colonies that have undergone plasmid excision in the second recombination, but it does not separate deletion mutants from wild-type revertants, and it does not have a positive selection for plasmid excision.

6.5.3 SELECTION FOR PLASMID EXCISION WITH ANTISENSE SECY

Recently, pKOR1 and pIMAY suicide plasmids were developed for allelic exchange in *Staphylococcus*.[54,55] These two vectors carry an antisense *secY* RNA gene that enables positive selection for the second recombination event. *SecY* is part of the transmembrane component of the general Sec protein secretion system, which is highly conserved in Gram-positive bacteria. In *Staphylococcus*, *secY* is vital for growth and survival. Expression of *secY* antisense RNA inhibits *secY* expression and prevents colony formation on agar plates.[110] *SecY* has not been characterized in *L. monocytogenes*, but transmembrane components *secY*EGDF are encoded in the genomes of *L. monocytogenes* strains EGDe (serotype 1/2a), F2365 (4b), F6854 (1/2a), and H7858 (4b).[56]

In pIMAY, the antisense *secY* RNA gene is under the control of a tetracycline-inducible promoter. Therefore, induction of *secY* antisense RNA expression in the presence of anhydrotetracycline (ATc) prevents the growth of bacteria that retain the integrated plasmid, thus providing a positive selection for chromosomal excision and loss of the plasmid. Our laboratory attempted to use these two plasmids to construct in-frame deletions in *L. monocytogenes*, but our results were unsuccessful because plasmid transformation/chromosomal integration (the first recombination event) occurred with very low efficiency in *L. monocytogenes* (data not shown).

6.5.4 CONSTRUCTION OF NEW SUICIDE PLASMID FOR *L. MONOCYTOGENES*

Although several plasmids have been used in *L. monocytogenes* for generating mutants by allelic exchange, construction of *L. monocytogenes* mutants has been challenging task due to lack of useful selection markers for the first and second recombination events.[57] To address this problem, a new suicide plasmid (pHoss1) was constructed that combines the pMAD backbone and the *secY* antisense cassette from pIMAY.[57] In pHoss1, the pMAD backbone provides efficient transformation and plasmid integration (the first recombination event) in *L. monocytogenes*, and expression of the *secY* antisense RNA provides an effective selection for plasmid excision (the second allelic exchange event) and generation of a markerless deletion.

Construction of the suicide plasmid pHoss1 is summarized in Figure 6.1. Briefly, a 1371 bp fragment carrying tetracycline-inducible antisense *secY* RNA gene was PCR amplified from pIMAY. The amplicon was digested with BseRI and BglII restriction enzymes and cloned into pMAD digested with the same two enzymes. Presence of the antisense *secY* RNA gene cassette in pMAD was confirmed by PCR and sequencing. The pHoss1 plasmid is 8995 bp. It contains a heat-sensitive origin of replication, four unique restriction sites (SalI, EcoRI, SmaI, and NcoI), erythromycin resistance gene, and a 1371 bp fragment encoding an antisense *secY* RNA gene driven by an inducible Pxyl/tetO promoter (Fig. 6.1).

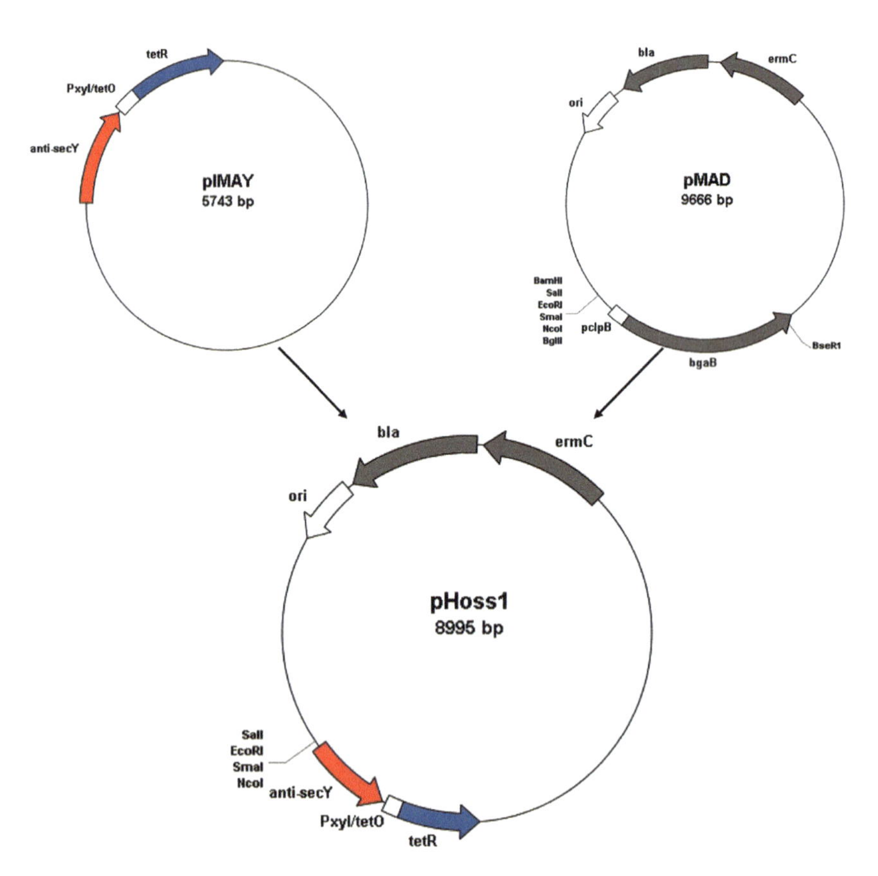

FIGURE 6.1 Construction of the suicide plasmid pHoss1. pHoss1 is derived from pMAD. The β-galactosidase gene (*bgaB*) in pMAD was deleted and replaced with tetracycline-inducible antisense *secY* gene (anti-*secY*) amplified from pIMAY. *ermC*, Ery resistance gene; *bla*, beta-lactamase; ori, the pBR322 origin of replication.

6.5.5 SELECTED EXAMPLE FOR IN-FRAME GENE DELETION IN *L. MONOCYTOGENES* WITH PHOSS1

To assess the usefulness of pHoss1 in *L. monocytogenes*, we constructed in-frame deletions of the *ispG* and *ispH* genes (LMOf2365_1460 and LMOf2365_1470). *IspG* and *IspH* are iron sulfate enzymes involved in catalyzing the final two steps in the mevalonate-independent (MEP) pathway for isoprenoid biosynthesis.[58,59] Enzymes in this pathway are encoded by six genes (*ispC, ispD, ispE, ispF, ispG,* and *ispH*) and result in the production of isopentenyl pyrophosphate (IPP) or its isomer dimethylallyl pyrophosphate (DMAPP).[60,61] 1-hydroxy-2-methyl-2-(E)-butenol 4-diphosphate synthase (*ispG*) and 4-hydroxy-3-methylbut-2-enyl diphosphate reductase (*ispH*) are the last two enzymes of this pathway. *IspG* catalyzes the conversion of 2C-methyl-D-erythritol 2,4-cyclodiphosphate (ME-2,4cPP) into 1-hydroxy-2-methyl-2-(E)-butenyl 4-diphosphate,[58] and *ispH* converts 1-hydroxy-2-methyl-2-(E)-butenyl 4-diphosphate into IPP and DMAPP.[59] Because the enzymes of the MEP pathway are not found in humans, this pathway has been used as a drug target for treatment of some bacterial infections and malaria.[62,63] These two genes are not possessed by nonpathogenic strains of *L. innocua* and *L. monocytogenes.*[64,65] Thus, we hypothesized that *ispG* and *ispH* genes could be essential for *L. monocytogenes* pathogenesis, and we tested this by developing mutants.

L. monocytogenes strain F2365 was used in this study; this strain was isolated from Mexican-style soft cheese and implicated in a 1985 outbreak of listeriosis in California.[20] Two *L. monocytogenes* mutant strains, *Lm*f2365Δ*ispG* and *Lm*f2365Δ*ispH* were constructed by in-frame deletions using the newly developed plasmid pHoss1 (Fig. 6.2A). Briefly, approximately 1 kb amplicons upstream and downstream of each gene were amplified separately with A and B (A/B) and C and D (C/D) primers listed in Table 6.2. Then overlap extension PCR was used to generate approximately 2 kb gene deletion amplicons,[66] which were cloned into pHoss1. The resulting plasmids were transformed into *L. monocytogenes* strain F2365 by electroporation, and the bacteria were recovered in BHI broth for 3 h at 30°C before spreading on BHI agar containing Ery (10 μg/mL).

We determined that *L. monocytogenes* is sensitive to more than 2.00 μg/mL ATc. Thus, we used 1.5 μg/mL ATc for induction of *secY* antisense RNA. A two-step procedure was used for replacement of the wild type *ispG* and *ispH* genes. In the first step, one colony was picked from a BHI agar plate with Ery into 2 mL BHI broth and grown overnight at 30°C, then streaked on a BHI agar plate with Ery and incubated at 42°C for 2 days, which was repeated twice. In the second step, Ery resistant colonies were grown overnight in BHI broth (no Ery) at 30°C, which was repeated twice. Cultures were passed in fresh BHI broth and grown at 42°C for 8 h, and mutants were selected by spreading diluted culture on a BHI agar plate containing 1.5 μg/mL ATc and incubating plates at 30°C for 3 days. Finally, 20 colonies were picked, and colony PCR was performed using A and D primers. Ery sensitivity was also checked. Confirmed deletion mutants were designated *Lm*f2365Δ*ispG* and *Lm*f2365Δ*ispH*.

TABLE 6. 2 Primers Used to Generate and Verify In-Frame Deletions.

Primers		Sequence (5'→3')[a]	RE[b]
Lm-ispG-F01	A	aagtcgactagcctaccatgctcctgaaa	SalI
Lm-ispG-R847	B	catagagaccgctcctttag	
Lm-ispG-F801	C	ctaaaggagcggtctctatgagataatcgtatcggggttt	
Lm-ispG-R01	D	aaccatggatggtaggaagtgatgcgagt	NcoI
Lm-ispH-F01	A	aagtcgaccgctaaataaggctgtgaacc	SalI
Lm-ispH-R900	B	tccgtagcaataaccacgag	
Lm-ispH-F900	C	ctcgtggttattgctacggagctaaaaaccgagcaactcct	
Lm-ispH-R01	D	aaccatggtccgtttctatatcggccaac	NcoI
Anti_secY F01		aagaggaggatctaatgattcaaacccttgtg	BseRI
Anti_secY R01		aaagatcttgaagttaccatcacggaaaaagg	BglII

[a]Underlined bases in primer C indicate reverse complemented primer B sequences.

[b]RE stands for restriction enzyme recognition sequence added to the 5' end of the primer sequence.

Using the pHoss1 plasmid, we were able to construct two *L. monocytogenes* mutant strains, *Lm*f2365Δ*ispG* and *Lm*f2365Δ*ispH* (Fig. 6.2B). Selection for the second recombination event was very efficient with pHoss1. Of the twenty potential *ispG* ATc-resistant mutant colonies screened by PCR, sixteen had deletion mutations, and four were wild-type revertants. Of the 19 potential *ispH* mutants screened, 100% were deletion mutants. Thus, at least 80% of the ATc-resistant colonies were deletion mutants, which confirm that antisense *secY* selection is very efficient in eliminating *L. monocytogenes* carrying the pHoss1 backbone. All positive mutants were also sensitive to Ery, confirming the loss of the pHoss1 plasmid. The resulting *Lm*f2365Δ*ispG* strain contained a deletion of 1098 bp from the *ispG* gene (99% of the ORF), and the *Lm*f2365Δ*ispH* strain contained a deletion of 948 bp from the *ispH* gene (95% of the ORF).

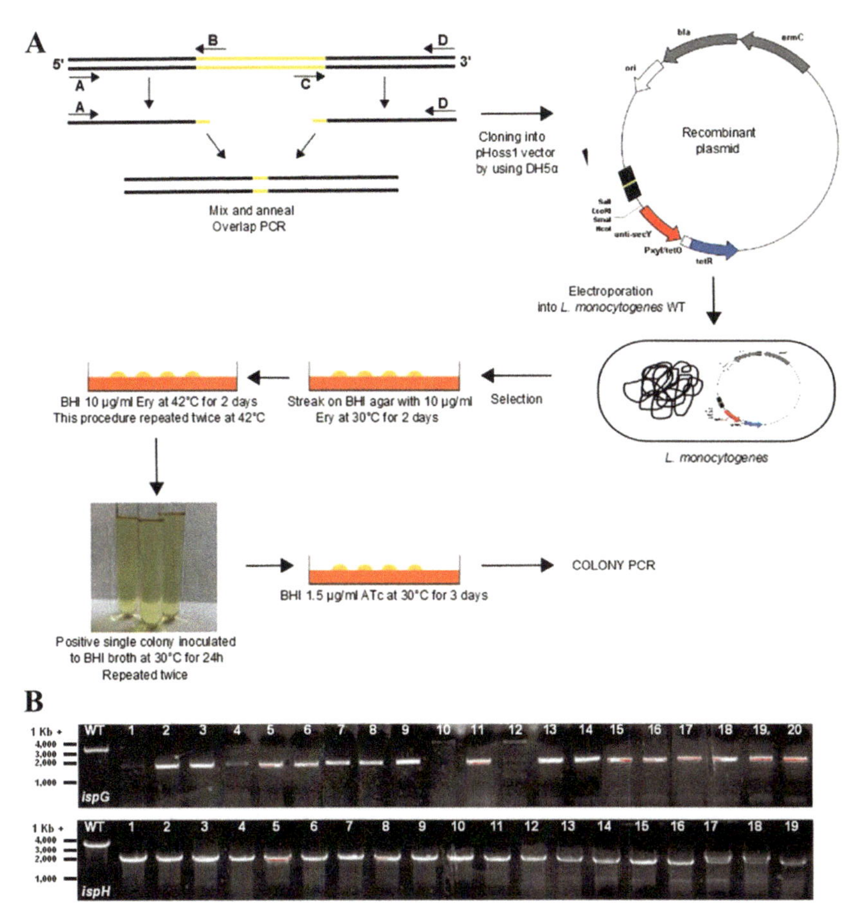

FIGURE 6.2 (A) Strategy used to create a deletion of *ispH* and *ispG* in *L. monocytogenes* F2365 using suicide plasmid pHoss1 and allelic exchange.[10] (B) PCR verification of Δ*ispG* and Δ*ispH* deletion using A and D primers. The sizes of 1 Kb Plus DNA Ladder (Life Technologies) bands are indicated on the left. WT indicates PCR fragment amplified from wild-type *L. monocytogenes* F2365. Numbers at the top are ATc-resistant listerial colonies picked randomly for PCR screening. 16/20 (80%) of *ispG* and 19/19 (100%) of *ispH* colonies showed the gene deletion fragment only. In *ispG*, four colonies showed wild-type revertants or merodiploids (lanes 1, 4, 10, and 12).

6.6 PATHOGENESIS

The main route of *Listeria* transmission is through ingestion of contaminated food products.[67,68] Other modes of transmission including transplacental from mother to fetus during pregnancy and via the birth canal during birth.[69,70] Transmission of the infection to farmers and veterinarians can occur through direct contact with diseased domestic farm animals during the birthing of animals.[71] Foods can become contaminated with *L. monocytogenes* in the produce growing environment, during processing, or during handling and preparation. Pasteurization and sufficient cooking kill *Listeria*; however, contamination may occur after cooking and before packaging. *L. monocytogenes* can multiply during the storage process in contaminated refrigerated food, which makes this pathogen a particular problem for food safety. For example, soft cheese is identified as one of the biggest high-risk foods and one of the leading causes of listeriosis due to the use of raw milk and because of replication and additional contamination during refrigeration.[10,72]

The infective dose is undetermined but is believed to vary with the strain, host susceptibility, and food source.[12] In a healthy person, 10–100 million viable bacteria or colony forming units "CFU" must be ingested to cause illness, while as few as 0.1–10 million CFU can cause illness in at-risk people.[12,73] Potential contamination sources include food workers, incoming air, raw materials, and food-processing environments. Among those, post-processing contamination at food-contact surfaces poses the greatest threat to products.[74]

L. monocytogenes can invade and reproduce within a wide range of mammalian cell types, including intestinal, hepatic, and neural cells. During foodborne infections, the bacteria invade enterocytes and translocate to the lymphatic system to reach target tissues (liver and spleen).[75] *L. monocytogenes* replicates in the liver and can then potentially spread to the brain or the placental barrier in pregnant women.[76] *L. monocytogenes* can survive and replicate in macrophages, which can contribute to dissemination. More than 50 listerial virulence factors have been described.[77]

6.7 IDENTIFICATION AND DETECTION

Historically, isolation of *L. monocytogenes* from food or clinical samples was based on its ability to grow at low temperatures. This method required incubation for prolonged periods at 4°C until the formation of visible colonies. This method takes up to several weeks and does not consistently detect the pathogen.[78] In the food industry, the FDA bacteriological and analytical method (BAM),[79] International Organization of

Standards (ISO) 11290, the USDA protocol (used for meat, eggs, poultry, and environmental samples),[80] and the Association of Analytical Chemists (AOAC/IDF) method 993.12 (for dairy samples)[81] are commonly used for discovery of *Listeria* species.

Listeria are Gram-positive, facultatively anaerobic, non-spore forming rods. Biochemical and phenotypic markers used for identification of *Listeria* species include the esculinase reaction and β-D-glucosidase activity. *Listeria* are also catalase-positive, oxidase-negative, and fermentative in sugars, which is detectable using commercial API *Listeria* (bioMerieux, France) and Micro-IDTM (Remel, USA). The API *Listeria* test (bio-Merieux) comprises 10 biochemical tests that allow differentiation in a microtube format (Table 6.3).[82] Another useful confirmatory test for *Listeria* species is the Christie–Atkins–Munch–Peterson (CAMP) test, which tests for hemolysis enhancement on sheep blood agar in the presence of other hemolytic bacteria. The hemolytic *Listeria* species are *L. monocytogenes*, *L. seeligeri*, and *L. ivanovii*; of these, *L. monocytogenes* and *L. seeligeri* are CAMP positive, and *L. ivanovii* is CAMP negative.[83]

Chromogenic substrates[84] incorporated into plating media allow direct identification of colonies based on specific biochemical characteristics. Some of these are commercially available. For example, media based on detection of phosphatidylinositol-phospholipase C (PI-PLC) and, to a lesser extent, phosphatidylcholine-phospholipase C (PC-PLC) allow direct detection and enumeration of pathogenic *Listeria* species (*L. monocytogenes*, *L. seeligeri*, and *L. ivanovii*). Several selective chromogenic agar media specifically designed for differentiation of *L. monocytogenes* are also available. Examples include bioMérieux ALOA® One Day agar, Oxoid Chromogenic *Listeria* agar (OCLA), and CHROMagar™ *Listeria*.

Immunoassay methods available include immunomagnetic separation and enzyme-linked immunosorbent assays (ELISA) incorporating fluorescent or colorimetric detection. These assays are based on *Listeria*-specific antibodies and are popular in food testing because of their simplicity, sensitivity, and accuracy. Also, testing can be conducted directly from enrichment media without sample preparation.[78] Recently, the use of chromogenic media in combination with immunomagnetic separation was investigated as an alternative method for detection of *L. monocytogenes* from foods.[85,86] Serotyping has relatively low discriminatory power

TABLE 6.3 Identification Scheme of *Listeria* Species on the API *Listeria* System as Provided by the Manufacturer.[82]

	L. monocytogenes	***L. innocua***	***L. seeligeri***	***L. ivanovii* subsp. *ivanovii***	***L. ivanovii* subsp. *londoniensis***	***L. welshmeri***	***L. grayi***
DIM	–	+	+	V	V	V	+
Esculin hydrolysis	+	+	+	+	+	+	+
α-Mannosidase	+	+	–	–	–	+	V
D-Arabitol	+	+	+	+	+	+	+
D-Xylose	–	–	+	+	+	+	–
L-Rhamnose	+	V	–	–	–	V	–
α-Methyl-D-glucoside	+	+	+	+	+	+	V
D-Ribose	–	–	–	+	–		+
Glucose-1-phosphate	–	–	–	+	V	–	–
D-Tagatose	–	–	–	–	–	+	–
Hemolysis	+	–	+	+	+	–	–
CAMP test (*S. aureus*)	+	–	+	–	–	–	–

+: positive reaction; –: negative reaction; V: variable reaction.

compared to molecular methods such as ribotyping or pulsed-field gel electrophoresis (PFGE).[87] Of the 13 *L. monocytogenes* serotypes, over 98% of isolates from human listeriosis belong to only four serotypes: 1/2a, 1/2b, 1/2c, and 4b.[88] Serotype 1/2a represents over 50% of the *L. monocytogenes* isolates recovered from foods and the environment, while serotype 4b strains are the most responsible for outbreak and sporadic case of human listeriosis.[89–92]

DNA hybridization is used for detection of *Listeria* in foods. It is applied using a labeled oligonucleotide probe that is complementary to target sequences; it requires comparatively large amounts of target DNA or RNA. Ribotyping relies on the separation and analysis of specific well-conserved DNA fragments. This method is often used in combination with serotyping to identify and trace a specific outbreak-associated strain of *L. monocytogenes* to a food source or to link seemingly unrelated listeriosis cases. Wiedmann et al.[93] separated *L. monocytogenes* into three lineages by ribotyping and PCR-restriction fragment length polymorphism of three virulence genes (*hyl*, *actA*, and *inlA*).

PCR is a very sensitive detection method capable of detecting small amounts of DNA, and it is established as a reliable and reproducible technique for differentiation of *L. monocytogenes* and other *Listeria* species. PCR assays to identify *Listeria* species other than *L. monocytogenes* have been developed.[94–97] Multiplex assays have been developed that are capable of differentiating *L. monocytogenes* from other *Listeria* species or subtyping within *L. monocytogenes.*[98–101] Analysis of differences in genomic content among strains of different serotypes and lineages led to the development of a PCR-based method that allows distinction of four *L. monocytogenes* subgroups: 1/2a-3a, 1/2c-3c, 1/2b-3b-7, and 4b-4d-4e.[102,103] However, it does not distinguish serotype 1/2a from 3a, 1/2c from 3c, 1/2b from 3b and 7, or 4b from 4d and 4e. Thus, there is still need for the development of a rapid PCR method that can distinguish high-risk *Listeria* serotypes 1/2a, 1/2b, 1/2c, and 4b from the low-risk serotypes to enable studies on the epidemiology of foodborne disease and the development of preventive strategies.

Recently, matrix-assisted laser desorption ionization-time of flight mass spectrometry (MALDI-TOF MS) has emerged as a valuable method for microbial identification and diagnosis.[104,105] This technique relies on the principle that proteins extracted from bacteria exhibit unique and representative spectral patterns that can be compared to a reference strain to identify bacterial species or subspecies.[106] MALDI-TOF MS methods for identification of *L. monocytogenes*, *L. innocua*, *L. ivanovii*, *L. seeligeri*, *L. welshimeri*, and *L. grayi* isolates were developed.[107] Putative "specific-identifying" biomarker ions were determined for *Listeria* species by averaging the results for several strains tested in multiple experiments, and the most abundant and conserved peaks were based mainly on detection of ribosomal protein fractions of each *Listeria* species. Thus, MALDI-TOF MS method has good potential for *Listeria* identification and control of preventive strategies in the food industry.

6.8 CONCLUSION AND FUTURE RESEARCH

The foodborne pathogen *Listeria monocytogenes* has been responsible for many disease outbreaks and food recalls. Listeriosis has one of the highest case fatality rates of the foodborne diseases. In this chapter, we provide background on *L. monocytogenes* and listeriosis. Current methods for genetic manipulation of *L. monocytogenes* and detection of the pathogen are described. In particular, a newly developed plasmid, pHoss1, is described, which has the potential to accelerate pathogenesis research of *L. monocytogenes* by enabling deletion of genes efficiently. New developments in molecular detection methods, including multiplex PCR and MALDI-TOF MS, have the potential for improving sensitivity and accuracy of detection in food and the environment. Improvements have been made in subtyping *L. monocytogenes* strains using molecular methods to allow better prediction of risk, but there is a need for more work to improve these methods further.

KEYWORDS

- ***Listeria monocytogenes***
- **pathogenesis**
- **homologous recombination**
- **bacterial detection**

REFERENCES

1. Buzby, J. C.; Roberts, T.; Jordan lin, C. T.; Macdonald, J. M. *Bacterial Foodborne Disease: Medical Costs & Productivity Losses;* Food and Consumer Economics Division, Economic Research Service, U.S. Department of Agriculture: Washington, DC, 1996.
2. Seeliger, H. P. R.; Jones, S. Genus *Listeria* Pirie, 1940, 383. In *Bergey's Manual of Systemic Bacteriology;* Sneath, P. H. A., Mair, N. S., Sharpe, M. E., Holt, J. G., Eds.; Williams and Wilkins: Baltimore, MD, 1986; Vol. 2, pp 1235–1245.
3. Gray, M. L.; Killinger, A. H. *Listeria monocytogenes* and Listeric Infections. *Bacteriol. Rev.* **1966,** *30,* 309–382.
4. Collins, M. D.; Wallbanks, S.; Lane, D. J.; Shah, J.; Nietupski, R.; Smida, J., et al. Phylogenetic Analysis of the Genus *Listeria* Based on Reverse Transcriptase Sequencing of 16S rRNA. *Int. J. Syst. Bacteriol.* **1991,** *41,* 240–246.
5. Schuchat, A.; Swaminathan, B.; Broome, C. V. Epidemiology of Human Listeriosis. *Clin. Microbiol. Rev.* **1991,** *4,* 169–183.
6. Ramaswamy, V. C. V.; Rejitha, J. S.; Lekshmi, M. U.; Dharsana, K. S.; Prasad, S. P.; Vijila, H. M. *Listeria*--Review of Epidemiology and Pathogenesis. *J. Microbiol. Immunol. Infect.* **2007,** *40* (1), 4–13.
7. Farber, J. M.; Peterkin, P. I. *Listeria monocytogenes,* A Food-Borne Pathogen. *Microbiol. Rev.* **1991,** *55,* 476–511.
8. Duche, O.; Tremoulet, F.; Glaser, P.; Labadie, J. Salt Stress Proteins Induced in *Listeria monocytogenes. Appl. Environ. Microbiol.* **2002,** *68,* 1491–1498.
9. Smerdon, C. A.; Woodward, D. L.; Young, C.; Rodgers, F. G.; Wiedmann, M. Correlations between Molecular Subtyping and Serotyping of *Listeria monocytogenes. J. Clin. Microbiol.* **2001,** *39,* 2704–2707.
10. Allerberger, F.; Wagner, M. *Listeriosis:* A Resurgent Foodborne Infection. *Clin. Microbiol. Infect.* **2010,** *16,* 16–23.
11. Erdenlig, S.; Ainsworth, A. J.; Austin, F. W. Production of Monoclonal Antibodies to *Listeria monocytogenes* and Their Application to Determine the Virulence of Isolates from Channel Catfish. *Appl. Environ. Microbiol.* **1999,** *65,* 2827–2832.
12. Bortolussi, R. Listeriosis: A Primer. *Can. Med. Assoc. J.* **2008,** *179,* 795–797.
13. Donnelly, C. W. *Listeria monocytogenes.* In *Guide to Foodborne Pathogens;* Abbé, R. G., Garcia, S., Eds.; John Wiley & Sons, Inc.: New York, NY, 2001; pp 99–132.
14. Buzby, J. C.; Roberts, T. The Economics of Enteric Infections: Human Foodborne Disease Costs. *Gastroenterology* **2009,** *136,* 1851–1862.
15. Centers for Disease Control and Prevention. Preliminary FoodNet Data on the Incidence of Infection with Pathogens Transmitted Commonly through Food—10 States, 2009. *MMWR Morb. Mortal. Wkly. Rep.* **2010,** *59,* 418–422.
16. Walderhang, M. *The Bad Bug Book: Foodborne Pathogenic Microorganisms and Natural Toxins Handbook*; US Food & Drug Administration, Center for Food Safety & Applied Nutrition: Washington, DC, 1992.
17. Rocourt, J.; Cossart, P. *Listeria monocytogenes.* In *Food Microbiology-Fundamentals and Frontiers;* Doyle, M. P., Beuchat, L. R., Montville, T. J., Eds.; American Society for Microbiology Press: Washington, DC, 1997; pp 337–352.

18. Ciesielski, C. A.; Hightower, A. W.; Parsons, S. K.; Broome, C. V. Listeriosis in the United States: 1980–1982. *Arch. Intern. Med.* **1988,** *148,* 1416–1419.
19. Gellin, B. G.; Broome, C. V.; Bibb, W. F.; Weaver, R. E.; Gaventa, S.; Mascola, L. The Epidemiology of Listeriosis in the United States--1986. Listeriosis Study Group. *Am. J. Epidemiol.* **1991,** *133,* 392–401.
20. Linnan, M. J.; Mascola, L.; Lou, X. D.; Goulet, V.; May, S.; Salminen, C., et al. Epidemic Listeriosis Associated with Mexican-Style Cheese. *N. Engl. J. Med.* **1988,** *319,* 823–828.
21. Centers for Disease Control and Prevention. Outbreak of *Listeria monocytogenes* Infections Associated with Pasteurized Milk from a Local Dairy--Massachusetts, 2007. *MMWR Morb. Mortal. Wkly. Rep.* **2008,** *57,* 1097–1100.
22. Food Safety Authority of Ireland (FSAI). The Control and Management of *Listeria monocytogenes* Contamination of Food. Food Safety Authority of Ireland: Ireland, 2007.
23. Aureli, P.; Fiorucci, G. C.; Caroli, D.; Marchiaro, G.; Novara, O.; Leone, L., et al. An Outbreak of Febrile Gastroenteritis Associated with Corn Contaminated by *Listeria monocytogenes. N. Engl. J. Med.* **2000,** *342,* 1236–1241.
24. Centers for Disease Control and Prevention. Incidence and Trends of Infection with Pathogens Transmitted Commonly Through Food—Foodborne Diseases Active Surveillance Network, 10 U.S. Sites, 2006–2013. *MMWR Morb. Mortal. Wkly. Rep.* **2013,** *62,* 283–287.
25. Centers for Disease Control and Prevention. Multistate Outbreak of Listeriosis Associated with Jensen Farms Cantaloupe--United States, August–September 2011. *MMWR Morb. Mortal. Wekly. Rep.* **2011,** *60,* 1357–1358.
26. Health. SMo. Infectious Disease Notifications in Singapore, 1990–2007. [Online]; 1990–2007. http://www.sgdi.gov.sg/
27. Seeliger, H. P.; Hohne, K. Serotyping of *Listeria monocytogenes* and Related Species. *Methods Microbiol.* **1979,** *13,* 31–49.
28. Cheng, Y. S. R.; Kathariou, S. Genomic Division/Lineages, Epidemic Clones and Population Structure. In *Handbook of Listeria monocytogenes;* Liu, D., Ed.; CRC Press: Boca Raton, FL, 2008; pp 337–357.
29. Nadon, C. A.; Woodward, D. L.; Young, C.; Rodgers, F. G.; Wiedmann, M. Correlations between Molecular Subtyping and Serotyping of *Listeria monocytogenes. J. Clin. Microbiol.* **2001,** *39,* 2704–2707.
30. Rasmussen, O. F, Skouboe, P.; Dons, L.; Rossen, L.; Olsen, J. E. *Listeria monocytogenes* Exists in at least Three Evolutionary Lines: Evidence from Flagellin, Invasive Associated Protein and Listeriolysin O Genes. *Microbiology* **1995,** *141* (9), 2053–2061.
31. Doumith, M.; Cazalet, C.; Simoes, N.; Frangeul, L.; Jacquet, C.; Kunst, F., et al. New Aspects Regarding Evolution and Virulence of *Listeria monocytogenes* Revealed by Comparative Genomics and DNA Arrays. *Infect. Immun.* **2004,** *72,* 1072–1083.
32. Roberts, A.; Nightingale, K.; Jeffers, G.; Fortes, E.; Kongo, J. M.; Wiedmann, M. Genetic and Phenotypic Characterization of *Listeria monocytogenes* Lineage III. *Microbiology* **2006,** *152,* 685–693.
33. 33. Ward, T. J.; Gorski, L.; Borucki, M. K.; Mandrell, R. E.; Hutchins, J.; Pupedis, K. Intraspecific Phylogeny and Lineage Group Identification Based on the prfA Virulence Gene Cluster of *Listeria monocytogenes. J. Bacteriol.* **2004,** *186,* 4994–5002.

34. den Bakker, H. C.; Bowen, B. M.; Rodriguez-Rivera, L. D.; Wiedmann, M. FSL J1-208, A Virulent Uncommon Phylogenetic Lineage IV *Listeria monocytogenes* Strain with a Small Chromosome Size and a Putative Virulence Plasmid Carrying Internalin-Like Genes. *Appl. Environ. Microbiol.* **2012,** *78,* 1876–1889.
35. Liu, D.; Lawrence, M. L.; Wiedmann, M.; Gorski, L.; Mandrell, R. E.; Ainsworth, A. J., et al. *Listeria monocytogenes* Subgroups IIIA, IIIB, and IIIC Delineate Genetically Distinct Populations with Varied Pathogenic Potential. *J. Clin. Microbiol.* **2006,** *44,* 4229–4233.
36. Tan, M. F.; Siow, C. C.; Dutta, A.; Mutha, N. V.; Wee, W. Y.; Heydari, H., et al. Development of *Listeria* Base and Comparative Analysis of *Listeria monocytogenes. BMC Genomics* **2015,** *16,* 755.
37. Glaser, P.; Frangeul, L.; Buchrieser, C.; Rusniok, C.; Amend, A.; Baquero, F., et al. Comparative Genomics of *Listeria* Species. *Science* **2001,** *294,* 849–852.
38. Nelson, K. E.; Fouts, D. E.; Mongodin, E. F.; Ravel, J.; DeBoy, R. T.; Kolonay, J. F., et al. Whole Genome Comparisons of Serotype 4b and 1/2a Strains of the Food-Borne Pathogen *Listeria monocytogenes* Reveal New Insights into the Core Genome Components of This Species. *Nucleic Acids Res* **2004,** *32,* 2386–2395.
39. Paul, D.; Steele, C. L.; Donaldson, J. R.; Banes, M. M.; Kumar, R.; Bridges, S. M., et al. Genome Comparison of *Listeria monocytogenes* Serotype 4a Strain HCC23 with Selected Lineage I and Lineage II *L. monocytogenes* Strains and Other *Listeria* Strains. *Genomics Data* **2014,** *2,* 219–225.
40. den Bakker, H. C.; Cummings, C. A.; Ferreira, V.; Vatta, P.; Orsi, R. H.; Degoricija, L., et al. Comparative Genomics of the Bacterial Genus *Listeria*: Genome Evolution is Characterized by Limited Gene Acquisition and Limited Gene Loss. *BMC Genomics* **2010,** *11,* 688.
41. Deng, X.; Phillippy, A. M.; Li, Z.; Salzberg, S. L.; Zhang, W. Probing the Pan-Genome of *Listeria monocytogenes*: New Insights into Intraspecific Niche Expansion and Genomic Diversification. *BMC Genomics* **2010,** *11,* 500.
42. Jaworski, D. D.; Clewell, D. B. A Functional Origin of Transfer (oriT) on the Conjugative Transposon Tn916. *J. Bacteriol* **1995,** *177,* 6644–6651.
43. Maloy, S. R.; Stewart, V. J.; Taylor, R. K. *Genetic Analysis of Pathogenic Bacteria: A Laboratory Manual* Cold Spring Harbor Laboratory Press: Cold Spring Harbor, NY, 1996.
44. Pritchett, M. A.; Zhang, J. K.; Metcalf, W. W. Development of a Markerless Genetic Exchange Method for *Methanosarcina acetivorans* C2A and Its Use in Construction of New Genetic Tools for Methanogenic Archaea. *Appl. Environ. Microbiol* **2004,** *70,* 1425–1433.
45. Kristich, C. J.; Manias, D. A.; Dunny, G. M. Development of a Method for Markerless Genetic Exchange in *Enterococcus faecalis* and Its Use in Construction of a srtA Mutant. *Appl. Environ. Microbiol.* **2005,** *71,* 5837–5849.
46. Reyrat, J. M.; Pelicic, V.; Gicquel, B.; Rappuoli, R. Counterselectable Markers: Untapped Tools for Bacterial Genetics and Pathogenesis. *Infect Immun* **1998,** *66,* 4011–4017.
47. Ried, J. L.; Collmer, A. An nptI-sacB-sacR Cartridge for Constructing Directed, Unmarked Mutations in Gram-negative Bacteria by Marker Exchange-eviction Mutagenesis. *Gene* **1987,** *57,* 239–246.

48. Maloy, S. R.; Nunn, W. D. Selection for Loss of Tetracycline Resistance by *Escherichia coli. J. Bacteriol* **1981,** *145,* 1110–1111.
49. Chakraborty, T.; Leimeister-Wachter, M.; Domann, E.; Hartl, M.; Goebel, W.; Nichterlein, T., et al. Coordinate Regulation of Virulence Genes in *Listeria monocytogenes* Requires the Product of the prfA Gene. *J. Bacteriol* **1992,** *174,* 568–574.
50. Li, G.; Kathariou, S. An Improved Cloning Vector for Construction of Gene Replacements in *Listeria monocytogenes. Appl. Environ. Microbiol* **2003,** *69,* 3020–3023.
51. Smith, K.; Youngman, P. Use of a New Integrational Vector to Investigate Compartment-Specific Expression of the *Bacillus subtilis* spoIIM Gene. *Biochimie* **1992,** *74,* 705–711.
52. Arnaud, M.; Chastanet, A.; Debarbouille, M. New Vector for Efficient Allelic Replacement in Naturally Nontransformable, Low-GC-Content, Gram-positive Bacteria. *Appl. Environ. Microbiol* **2004,** *70,* 6887–6891.
53. Wuenscher, M. D.; Kohler, S.; Goebel, W.; Chakraborty, T. Gene Disruption by Plasmid Integration in *Listeria monocytogenes:* Insertional Inactivation of the Listeriolysin Determinant lisA. *Mol. Gen. Genet* **1991,** *228,* 177–182.
54. Monk, I. R, Shah, I. M.; Xu, M.; Tan, M. W.; Foster, T. J. Transforming the Untransformable: Application of Direct Transformation to Manipulate Genetically *Staphylococcus aureus* and *Staphylococcus epidermidis MBio.* **2012,** *3,* e00277–11.
55. Bae, T.; Schneewind, O. Allelic Replacement in *Staphylococcus aureus* with Inducible Counter-selection. *Plasmid* **2006,** *55,* 58–63.
56. Desvaux, M.; Hebraud, M. The Protein Secretion Systems in *Listeria:* Inside Out Bacterial Virulence. *FEMS Microbiol. Rev* **2006,** *30,* 774–805.
57. Abdelhamed, H.; Lawrence, M. L.; Karsi, A. A Novel Suicide Plasmid for Efficient Gene Mutation in *Listeria monocytogenes. Plasmid* **2015,** *81,* 1–8.
58. Hecht, S.; Eisenreich, W.; Adam, P.; Amslinger, S.; Kis, K.; Bacher, A., et al. Studies on the Nonmevalonate Pathway to Terpenes: The Role of the GcpE (IspG) Protein. *Proc. Natl. Acad. Sci. USA* **2001,** *98,* 14837–14842.
59. Altincicek, B.; Duin, E. C.; Reichenberg, A.; Hedderich, R.; Kollas, A. K.; Hintz, M., et al. LytB Protein Catalyzes the Terminal Step of the 2-C-methyl-D-erythritol-4-Phosphate Pathway of Isoprenoid Biosynthesis. *FEBS Lett* **2002,** *532,* 437–440.
60. Rohmer, M. The Discovery of a Mevalonate-Independent Pathway for Isoprenoid Biosynthesis in Bacteria, Algae and Higher Plants. *Nat. Prod. Rep* **1999,** *16,* 565–574.
61. Hunter, W. N. The Non-Mevalonate Pathway of Isoprenoid Precursor Biosynthesis. *J. Biol. Chem* **2007,** *282,* 21573–21577.
62. Obiol-Pardo, C.; Rubio-Martinez, J.; Imperial, S. The Methylerythritol Phosphate (MEP) Pathway for Isoprenoid Biosynthesis as a Target for the Development of New Drugs against Tuberculosis. *Curr. Med. Chem* **2011,** *18,* 1325–1338.
63. Nakagawa, K.; Takada, K.; Imamura, N. Probable Novel MEP Pathway Inhibitor and Its Binding Protein, IspG. *Biosci. Biotechnol. Biochem* **2013,** *77,* 1449–1454.
64. Begley, M.; Bron, P. A.; Heuston, S.; Casey, P. G.; Englert, N.; Wiesner, J., et al. Analysis of the Isoprenoid Biosynthesis Pathways in *Listeria monocytogenes* Reveals a Role for the Alternative 2-C-methyl-D-erythritol 4-Phosphate Pathway in Murine Infection. *Infect Immun* **2008,** *76,* 5392–5401.

65. Steele, C. L.; Donaldson, J. R.; Paul, D.; Banes, M. M.; Arick, T.; Bridges, S. M., et al. Genome Sequence of Lineage III *Listeria monocytogenes* Strain HCC23. *J. Bacteriol* **2011,** *193,* 3679–3680.
66. Horton, R. M.; Cai, Z. L.; Ho, S. N.; Pease, L. R. Gene Splicing by Overlap Extension: Tailor-made Genes Using the Polymerase Chain Reaction. *Biotechniques* **1990,** *8,* 528–535.
67. Low, J. C.; Donachie, W. A Review of *Listeria monocytogenes* and Listeriosis. *Vet. J.* **1997,** *153,* 9–29.
68. Gahan, C. G.; Hill, C. Gastrointestinal Phase of *Listeria monocytogenes* Infection. *J. Appl. Microbiol* **2005,** *98,* 1345–1353.
69. Roberts, A. J.; Wiedmann, M. Pathogen, Host and Environmental Factors Contributing to the Pathogenesis of Listeriosis. *Cell. Mol. Life Sci* **2003,** *60,* 904–918.
70. Mylonakis, E.; Paliou, M.; Hohmann, E. L.; Calderwood, S. B.; Wing, E. J. Listeriosis during Pregnancy: A Case Series and Review of 222 Cases. *Medicine (Baltimore)* **2002,** *81,* 260–269.
71. Regan, E. J.; Harrison, G. A.; Butler, S.; McLauchlin, J.; Thomas, M.; Mitchell, S. Primary Cutaneous Listeriosis in a Veterinarian. *Vet. Rec.* **2005,** *157,* 207.
72. Schlech, W. F., 3rd.; Lavigne, P. M.; Bortolussi, R. A.; Allen, A. C.; Haldane, E. V.; Wort, A. J., et al. Epidemic Listeriosis--Evidence for Transmission by Food. *N. Engl. J. Med* **1983,** *308,* 203–206.
73. Heinitz, M. L.; Johnson, J. M. The Incidence of *Listeria* spp., *Salmonella* spp., and *Clostridium botulinum* in Smoked Fish and Shellfish. *J. Food Prot* **1998,** *61,* 318–323.
74. Walderhang, M. *The Bad Bug Book: Foodborne Pathogenic Microorganisms and Natural Toxins Handbook;* International Medical Publishing: McLean, VA, 2004.
75. Marco, A. J.; Altimira, J.; Prats, N.; Lopez, S.; Dominguez, L.; Domingo, M., et al. Penetration of *Listeria monocytogenes* in Mice Infected by the Oral Route. *Microb. Pathog.* **1997,** *23,* 255–263.
76. Vazquez-Boland, J. A.; Kuhn, M.; Berche, P.; Chakraborty, T.; Dominguez-Bernal, G.; Goebel, W., et al. *Listeria* Pathogenesis and Molecular Virulence Determinants. *Clin. Microbiol. Rev* **2001,** *14,* 584–640.
77. Camejo, A.; Carvalho, F.; Reis, O.; Leitao, E.; Sousa, S.; Cabanes, D. The Arsenal of Virulence Factors Deployed by *Listeria monocytogenes* to Promote Its Cell Infection Cycle. *Virulence* **2011,** *2,* 379–394.
78. Gasanov, U.; Hughes, D.; Hansbro, P. M. Methods for the Isolation and Identification of *Listeria* spp. and *Listeria monocytogenes:* A Review. *FEMS Microbiol. Rev* **2005,** *29,* 851–875.
79. Center for Food Safety, Applied Nutrition. Rapid Methods for Detecting Foodborne Pathogens. Bacteriological Analytical Manual Online January 2001, 1–14.
80. United States Department of Agriculture. Food Safety and Inspection Service. Isolation and Identification of *Listeria monocytogenes* from Red Meat, Poultry, Egg and Environmental Samples. *Microbiology Laboratory Guidebook;* Microbiology Division, Washington, DC, 1998.
81. Cunniff P, Association of Official Analytical Chemists. Official Methods of Analysis of AOAC International. Association of Official Analytical Chemists, Washington, DC, 1995.

82. Bille, J.; Catimel, B.; Bannerman, E.; Jacquet, C.; Yersin, M. N.; Caniaux, I., et al. API *Listeria,* A New and Promising One-day System to Identify *Listeria isolates. Appl. Environ. Microbiol* **1992,** *58,* 1857–1860.
83. Groves, R. D.; Welshimer, H. J. Separation of Pathogenic from Apathogenic *Listeria monocytogenes* by Three In Vitro Reactions. *J. Clin. Microbiol.* **1977,** *5,* 559–563.
84. Reissbrodt, R. New Chromogenic Plating Media for Detection and Enumeration of Pathogenic *Listeria* spp.--An Overview. *Int. J. Food Microbiol* **2004,** *95,* 1–9.
85. Hudson, J. A.; Lake, R. J.; Savill, M. G.; Scholes, P.; McCormick, R. E. Rapid Detection of *Listeria monocytogenes* in Ham Samples Using Immunomagnetic Separation Followed by Polymerase Chain Reaction. *J. Appl. Microbiol* **2001,** *90,* 614–621.
86. Wadud, S.; Leon-Velarde, C. G.; Larson, N.; Odumeru, J. A. Evaluation of Immunomagnetic Separation in Combination with ALOA *Listeria* Chromogenic Agar for the Isolation and Identification of *Listeria monocytogenes* in Ready-to-Eat Foods. *J. Microbiol. Methods* **2010,** *81,* 153–159.
87. Brosch, R.; Chen, J.; Luchansky, J. B. Pulsed-Field Fingerprinting of Listeriae: Identification of Genomic Divisions for *Listeria monocytogenes* and Their Correlation with Serovar. *Appl. Environ. Microbiol* **1994,** *60,* 2584–2592.
88. Swaminathan, B.; Gerner-Smidt, P. The Epidemiology of Human Listeriosis. *Microbes Infect* **2007,** *9,* 1236–1243.
89. Jacquet, C.; Gouin, E.; Jeannel, D.; Cossart, P.; Rocourt, J. Expression of ActA, Ami, InlB, and Listeriolysin O in *Listeria monocytogenes* of Human and Food Origin. *Appl. Environ. Microbiol* **2002,** *68,* 616–622.
90. Mead, P. S.; Dunne, E. F.; Graves, L.; Wiedmann, M.; Patrick, M.; Hunter, S., et al. Nationwide Outbreak of Listeriosis due to Contaminated Meat. *Epidemiol. Infect* **2006,** *134,* 744–751.
91. Tresse, O.; Shannon, K.; Pinon, A.; Malle, P.; Vialette, M.; Midelet-Bourdin, G. Variable Adhesion of *Listeria monocytogenes* Isolates from Food-processing Facilities and Clinical Cases to Inert Surfaces. *J. Food Prot* **2007,** *70,* 1569–1578.
92. Pan, Y.; Breidt, F., Jr.; Kathariou, S. Competition of *Listeria monocytogenes* Serotype 1/2a and 4b Strains in Mixed-culture Biofilms. *Appl. Environ. Microbiol* **2009,** *75,* 5846–5852.
93. Wiedmann, M.; Bruce, J. L.; Keating, C.; Johnson, A. E.; McDonough, P. L.; Batt, C. A. Ribotypes and Virulence Gene Polymorphisms Suggest Three Distinct *Listeria monocytogenes* Lineages with Differences in Pathogenic Potential. *Infect. Immun* **1997,** *65,* 2707–2716.
94. Liu, D.; Ainsworth, A. J.; Austin, F. W.; Lawrence, M. L. PCR Detection of a Putative N-Acetylmuramidase Gene from *Listeria ivanovii* Facilitates Its Rapid Identification. *Vet. Microbiol* **2004,** *101,* 83–89.
95. Liu, D.; Ainsworth, A. J.; Austin, F. W.; Lawrence, M. L. Identification of a Gene Encoding a Putative Phosphotransferase System Enzyme IIBC in *Listeria welshimeri* and Its Application for Diagnostic PCR. *Lett. Appl. Microbiol* **2004,** *38,* 151–157.
96. Liu, D.; Lawrence, M. L.; Ainsworth, A. J.; Austin, F. W. Species-specific PCR Determination of *Listeria seeligeri. Res. Microbiol* **2004,** *155,* 741–746.
97. Volokhov, D.; Rasooly, A.; Chumakov, K.; Chizhikov, V. Identification of *Listeria* Species by Microarray-based Assay. *J. Clin. Microbiol* **2002,** *40,* 4720–4728.

98. Herman, L. M.; De Block, J. H.; Moermans, R. J. Direct Detection of *Listeria monocytogenes* in 25 Milliliters of Raw Milk by a Two-step PCR with Nested Primers. *Appl. Environ. Microbiol* **1995,** *61,* 817–819.
99. Bansal, N. S.; McDonell, F. H.; Smith, A.; Arnold, G.; Ibrahim, G. F. Multiplex PCR Assay for the Routine Detection of *Listeria* in Food. *Int. J. Food Microbiol* **1996,** *33,* 293–300.
100. Liu, D.; Lawrence, M. L.; Austin, F. W.; Ainsworth, A. J. A Multiplex PCR for Species- and Virulence-Specific Determination of *Listeria monocytogenes. J. Microbiol. Methods* **2007,** *71,* 133–140.
101. Sauders, B. D.; Overdevest, J.; Fortes, E.; Windham, K.; Schukken, Y.; Lembo, A., et al. Diversity of *Listeria* Species in Urban and Natural Environments. *Appl. Environ. Microbiol* **2012,** *78,* 4420–4433.
102. Borucki, M. K.; Call, D. R. *Listeria monocytogenes* Serotype Identification by PCR. *J. Clin. Microbiol* **2003,** *41,* 5537–5540.
103. Doumith, M.; Buchrieser, C.; Glaser, P.; Jacquet, C.; Martin, P. Differentiation of the Major *Listeria monocytogenes* Serovars by Multiplex PCR. *J. Clin. Microbiol* **2004,** *42,* 3819–3822.
104. Bizzini, A.; Greub, G. Matrix-Assisted Laser Desorption Ionization Time-of-flight Mass Spectrometry, a Revolution in Clinical Microbial Identification. *Clin. Microbiol. Infect* **2010,** *16,* 1614–1619.
105. Schmidt, V.; Carosch, A.; Marz, P.; Sander, C.; Vacata, V.; Kalka-Moll, W. Rapid Identification of Bacteria in Positive Blood Culture by Matrix-assisted Laser Desorption Ionization Time-of-flight Mass Spectrometry. *Eur. J. Clin. Microbiol. Infect. Dis.* **2012,** *31,* 311–317.
106. Marvin, L. F.; Roberts, M. A.; Fay, L. B. Matrix-assisted Laser Desorption/Ionization Time-of-flight Mass Spectrometry in Clinical Chemistry. *Clin. Chim. Acta.* **2003,** *337,* 11–21.
107. Barbuddhe, S. B.; Maier, T.; Schwarz, G.; Kostrzewa, M.; Hof, H.; Domann, E., et al. Rapid Identification and Typing of *Listeria* Species by Matrix-assisted Laser Desorption Ionization-time of Flight Mass Spectrometry. *Appl. Environ. Microbiol.* **2008,** *74,* 5402–5407.
108. Bean, N. H.; Griffin, P. M.; Goulding, J. S.; Ivey, C. B. Foodborne Disease Outbreaks, 5-year Summary, 1983–1987. *MMWR CDC Surveill. Summ.* **1990,** *39,* 15–57.
109. Doyle, M. P.; Schoeni, J .L. Selective-enrichment Procedure for Isolation of *Listeria monocytogenes* from Fecal and Biologic Specimens. *Appl. Environ. Microbiol.* **1986,** *51,* 1127–1129.
110. Ji, Y.; Zhang, B.; Van, S. F.; Horn, W. P.; Woodnutt, G.; Burnham, M. K. R.; Rosenberg, M. Identification of Critical Staphylococcal Genes Using Conditional Phenotypes Generated by Antisense RNA. *Science* **2001,** *293,* 2266–2269.

CHAPTER 7

SHIGELLA: A THREAT TO THE FOOD INDUSTRY

SUSHMA GURUMAYUM[1], SUSHREE SANGITA SENAPATI[1], PRASAD RASANE[2,3*], and SAWINDER KAUR[3]

[1]*Department of Microbiology, College of Allied Health Sciences, Assam Down Town University, Panikhaiti, Guwahati 781026, India*

[2]*Centre of Food Science and Technology, Banaras Hindu University, Varanasi 221005, Uttar Pradesh, India*

[3]*Department of Food Technology and Nutrition, Lovely Professional University, Jalandhar 144411, Punjab, India*

[*]*Corresponding author. E-mail: rasaneprasad@gmail.com*

CONTENTS

Abstract 180
7.1 Introduction 180
7.2 Biological Characteristics 181
7.3 Epidemiology 182
7.4 Methods of Isolation and Identification 184
7.5 Pathogenesis 190
7.6 Food Items Associated with *Shigella* Outbreaks 191
7.7 Treatment and Prevention 196
7.8 HACCP for *Shigella* Outbreak Management 197
Keywords 197
References 197

ABSTRACT

Shigella has proved to be a great menace in causing foodborne illness. Numerous species of *Shigella* have been reported to cause diseases such as dysentery, seizures, toxemia, shigellosis, and neurotoxicity. It could be transmitted through various biotic and abiotic vectors making it a potential epidemic organism. Several methods of *shigella* detection such as molecular methods (multiplex PCR and ATP bioluminescence), immunological methods (ELISA) and biophysical and biochemical methods (electrical impedometry and flow cytometry) have been developed so far. Although several treatments and curative techniques have been suggested worldwide, there is a constant threat of drug resistance with *Shigella*. Food processing areas are major source of *Shigella* contamination hence certain hazard analysis programs followed in such facilities needs to be upgraded to curb *Shigella* contamination.

7.1 INTRODUCTION

The prevalence of *Shigella* menace in food industry is a major concern. There are different sources of *Shigella* exposure and contamination in raw materials and in methods of processing of food items. *Shigella dysenteriae*, *Shigella flexneri*, *Shigella boydii,* and *Shigella sonnei* are the four recognized species of *Shigella.* There are no known non-human reservoirs. Foodborne shigellosis accounts for majority of the cases of food poisoning incidences in the world. Main cause of *Shigella* exposure to food is fecal-oral route of transmission and poor hygiene. Some of the vehicle food items associated are raw fruits, raw vegetables, salads, chicken, shellfish, salad dressings, marinades, salsa, cheese, dips, and toppings. Major losses are incurred in food processing industries due to *Shigella* contamination and outbreaks. There are several conventional methods of detection of *Shigella* and many more recent techniques have been newly developed. These include use of molecular methods like polymerase chain reaction (PCR)-based methods, immunological techniques, and use of biosensors. Methods of handling, processing, novel methods of packaging, and distribution of fresh food produce need to be monitored to prevent *Shigella* related food poisoning and outbreaks. Hazard analysis and critical control point (HACCP) programs specific to *Shigella* related food contamination should be developed and followed to ensure complete food safety. Finally, legislation of new standards for *Shigella* monitoring in food industry is of utmost requirement to remove the threat of *Shigella* in food industry. This chapter gives a comprehensive report of *Shigella* contamination sources; types of strains responsible for food poisoning; different food vehicles; conventional and molecular detection methods and the HACCP plan and legislation of standards in different parts of the world for food safety will be covered.

7.2 BIOLOGICAL CHARACTERISTICS

The genus *Shigella* belongs to the family *Enterobacteriacae.* They are Gram-negative rods appear singly, in pairs and in chains measuring 0.3–1 μm in diameter and 1–6 μm in length. Their growth temperature ranges from 10 to 40°C (optima of 37°C) and the optimum pH is 7.4. *Shigella* are aerobes as well as facultative anaerobes and non-motile in nature because of the absence of flagella.[1] Depending on the presence of somatic antigen (O) the species of Shigella are divided into *S. dysenteriae* (15 serotypes), *S. flexneri* (6 serotypes which can be sub-divided into sub-serotypes), *S. boydii* (20 serotypes) and *S. sonnei* (1 serotype, 2 variants—rough and smooth). *S. dysenteriae* (15 serotypes) are the intestinal pathogens of humans and primates cause bacillary dysentery. Serotype 1 is responsible for more severe disease producing a potent exotoxin known as Shiga toxin. *S. flexneri* (8 serotypes) cause disease in few cases. The disease caused by *S. sonnei* (serotype 1) is milder than the other *Shigella* species.[2] Humans are the primary reservoir; the differences in the species can be identified depending on their biochemical characterization. *S. dysenteriae* and *S. sonnei* can ferment sugar without gas production. *S. sonnei, S. flexneri,* and *S. boydii* can ferment Mannitol. This organism is urease and oxidase negative as they do not produce the enzyme urease as well as oxidase. Species like *S. dysenteriae* (15 serotypes), *S. flexneri* (6 serotypes which can be sub-divided into sub-serotypes), *S. boydii* (20 serotypes) and *S. sonnei* (1 serotype, 2 variants—rough and smooth) are known for the human infection. *S. dysenteriae* is found in the intestinal tract of humans and primates.[3–8] *Shigella* spreads through oral and fecal route of contamination. Especially drinking water and food are the main source of contamination because it can survive in water at room temperature for 6 months. The flies are the vector for *Shigella* as they transmit from human fecal matter to foods. Infections through direct contact with the infected person are also common. Health and hygiene are also responsible for *Shigella* contamination as *S. dysenteriae* serotype 1 can survive on human skin up to 1 h and a very small amount of it can cause severe infection. Gut of the patient is the main reservoir of *Shigella* in the world.[2,6,8,9] The *Shigella* species distribution is different worldwide.[2,6,8] Table 7.1 shows the reported global distribution of different species.

TABLE 7.1 Reported Global Distribution of Different *Shigella* Species.

Organisms	Distribution
S. dysenteriae	Densely populated areas of South America, Africa, and Asia
S. dysenteriae (Serotype 1)	India, Malaysia, and Guatemala
S. dysenteriae (Serotype 2)	Yemen and Nigeria
S. flexneri	Found in areas where endemic shigellosis occurs
S. boydii	Occurs sporadically
S. sonnei usually	Western developed countries such as France and United States

7.3 EPIDEMIOLOGY

Shigellosis is an endemic and sporadic disease caused by *Shigella*. The *Shigella* outbreak is found mostly in developing countries. The outbreak of the disease is more in crowded population. Poor hygienic condition is also responsible for this infection. The disease however, is a global threat.

It is estimated that 164.7 million are suffered from *Shigella* infection each year. Among them 163.2 million cases are reported in developing countries. From 163.2 million, 1.1 million death occur because of the infection. The children under 5 years are more infected, that is, 61% of all deaths.[2,6,10] In Bangladesh, during 1975–1985 children less than 5 years were reported dead due to acute or chronic watery diarrhea and dysentery.[11,12] In Central America 112,000 people suffered from *Shigella* infection and out of which 13,000 died in 1970. Five thousand people of Texas, United States reported to be infected in 1985. During May–June 1994, the European countries were reported with *S. sonnei* infection. Similar reports were in Paris in 1996.[2,4]

7.3.1 FOODBORNE SHIGELLOSIS

Shigellosis is an acute enteric infection caused by four known species of *Shigella,* namely, *S. flerneri, S. dysenteriae, S. sonnei,* and *S. boydii* characterized by nausea, vomiting, tenesmus, and bloody stools. The symptoms begin 1–4 days from infections and last for about a week. Though all the four species are capable of causing severe disease, infection caused by *S. flerneri* is usually involved in causing life-threatening form of dysentery. However, mild infections of *Shigella* often go undiagnosed and are not treated. In minors, elderly and in immune-compromised conditions infections may cause seizures, toxemia, and neurotoxicity. *S. sonnei* infections often do not result into severe diseased condition, while *S. dysenteriae* type 1 produces a toxin known as Shiga toxin resulting in intestinal perforations, mega colon intoxication and hemolytic uremic syndrome.[13] In rare infectious conditions individuals with HLA-B27 antigen may develops certain form of arthritis. Several outbreaks have been detected so far due to consumption of *Shigella* contaminated foods such as fresh produce, namely, lettuce,[14] sugar peas, and processed foods such as porridges. Infections are reported worldwide, namely, China,[12] Hawaii, Denmark, India, and South Africa.[13] *Shigella* spp. are highly infectious and are capable of causing infection with load as small as 10–200 cells.[15] If proper hygiene is not followed, *Shigella* infections are most likely. In various populated places such as children day-care and schools, infections are common amongst caretakers and children. Children of 1–4 years are found to be high at risk. Infections of about 12.5–28.3 cases per 100,000 children per year are reported.[16] Statistics also suggest that infections are also gender based, with females are at greater risk than males.[17] Surveillance reports suggest that apparently amongst young population aged 20–29 years of age about 70–75% of infections due to *S. sonnei* are reported in females. Overall about 51–53% infections are reported amongst females. About 30% of infections are reported due to secondary infections amongst children. *Shigella* infections can also be transmitted. Other modes of infection are due to contaminated pool water, population exposed to unhygienic religious places and homosexual practices.[16–19]

Foodborne shigellosis is not seasonal.[20] Though large amount of infections are caused due to secondary infections, foodborne infections are not uncommon. About 130,000 cases of foodborne shigellosis have been reported each year in United States itself.[21] Although surveillance data in developing countries like India and South Africa are limited, food borne shigellosis is reported in these countries, affecting global food trade. Food based infections are commonly transmitted through settings such as restaurants, schools, prisons and during festivals. Commercial unhygienic settings producing foods are mostly responsible for these infections. Cases not associated with outbreaks are regarded as sporadic accounts to majority of infections. Near about one-third of sporadic cases are attributed to foodborne infections.[22]

7.3.2 MOLECULAR EPIDEMIOLOGY

There has been increasing interest in the study of molecular epidemiology of genetically encoded virulence factors and antimicrobial resistance markers of *Shigella.* These information are helpful to understand transmission mechanisms and patterns, antibiotic resistance and severity of outbreaks. Several virulence factors have been reported in *Shigella*. Virulent genes of *Shigella* may be present in isolates from both symptomatic and asymptomatic people. Genes associated with virulence include virulence regulator (*VirF*), secreted autotransporter toxin (*sat*), *Shigella* enterotoxin 1 subunit A (*setA*), *Shigella* enterotoxin 1 subunit B (*setB*), *Shigella* enterotoxin 2 (*sen*) and epithelial cell penetration encoded by invasion associated locus (*ial*).[23]

Shigella has adapted well too many antimicrobial agents due to its ability to carry mobile genetic elements that may facilitate inter- and intra-species dissemination of antimicrobial resistance genes. Class 1 integron is the molecular virulence gene in *S. flexneri.*[24]

The main mechanism of quinolone resistance involves accumulation of sequential mutations in DNA gyrase and DNA topoisomerase IV. The plasmid-mediated quinolone resistance (PMQR) has been described due to mutations in the *aac(60)-Ibcr* gene that encodes for a variant of aminoglycoside acetyltransferase, also known to reduce ciprofloxacin activity.[25] PMQR genes in shigellae have been reported from United States, Japan, China, and India.[26,27]

7.4 METHODS OF ISOLATION AND IDENTIFICATION

Pathogens are ubiquitous reaching all prospects of life starting from food, water, soil, etc. Food and water are the main source of enteric pathogens like *Shigella* that are responsible for global epidemics. Their presence can cause severe threats to the human health. Therefore, the current trends in food and dairy industry follow safety measures before food materials reach the consumers. Food microbiologists play a major role in maintaining food safety measures.

To satisfy consumers demand for safe food supply, many conventional and recent techniques have been introduced. The traditional route of microbial analysis in food sample follows conventional methods which include pre-enrichment, selective enrichment, selective plating, biochemical characterization, and serological confirmation.[28] Qualitative and quantitative analysis of pathogens present in the food sample is possible by these methods. The process of identification of a particular food pathogen through conventional methods is very susceptible and cheap.[29] In traditional methods identification of food pathogen are confirmed by a complete set of tests.[30] These methods of identification are time consuming as it depends on the growth of the particular bacteria and at the same time the methods are laborious as it involves culture media preparation, biochemical tests, etc.[31,32]

The microbial analysis of fruit samples was performed in Karnataka, India. Samples were collected in sterile polythene bags which have been maintained in an aseptic condition in the laboratory and homogenized and filtered using Whatmann no 1 filter paper. To isolate the microorganisms, filtrate was streaked on differential media such as EMB, XLD, and MacConkey agar. The pure culture of each isolate was obtained and identified by cultural, morphological and biochemical characteristics (indole, methyl red, TSI, nitrate reductase test, citrate test, gelatin, catalase, and oxidase tests).[33]

Study on isolation of *Shigella* from food samples including 15 milk powder, 6 horlicks, 6 honey, 6 chutney, 6 chocolates, 10 biscuits, and 5 spices from retail stores was carried out in Dhaka, Bangladesh. Pathogens in these food items were detected using specific culture media known as Enterobacteriaceae Enrichment Broth (EEB). For further identification of bacteria present in these samples different culture media such as violet red bile glucose agar (VRBG), MacConkey agar (HiMedia, India), Xylose lysine deoxycholate (XLD) agar (Oxoid, UK), and *Cronobacter sakazakii* agar (HiMedia, India) were used. The isolated colonies were then subjected to grow on tryptone soy agar (TSA) for pigment production. The identification procedure for specific strain includes different biochemical tests like oxidase test, catalase test, citrate utilization test, methyl red, Voges–Proskauer, Kligler's iron agar (KIA), nitrate reduction test, arginine decarboxylase, gelatin hydrolysis and indole production, production of gas from glucose, and carbohydrate fermentation assay (xylose, trehelose, ducitol, arabinose, salicin, mannitol, sucrose, lactose, sorbitol, maltose, and esculin). Identification was done according to the Bergey's manual of systematic bacteriology.[34]

In Seoul, Korea, 92 varieties of food samples including ground beef, pork, chicken, onions, celery, lettuce, cabbage, spinach, parsley, cucumbers, potatoes, bell peppers, broccoli, cheese, shrimp, short-necked clams, manila clams, and sea cucumbers were collected from local retail markets to identify *Shigella* using a newly designed selective medium (HEX medium, modified Hektoen Enteric agar medium). Lactose, sucrose, D-xylose, and salicin as a differentiation marker were the main constituents and the concentration of bile salts No.3 of 0.3% helped in inhibition of Gram-positive bacteria. The growth of *S. flexneri, S. dysenteriae, S. boydii,* and *S. sonnei* on HEX medium was successful and green color pigmentation was observed; however, Hafnia alvei, was false positive for *Shigella* on HE agar as it was observed as orange colonies on HEX medium.

7.4.1 COMPOSITION OF SHIGELLA DETECTION MEDIUM: (MODIFIED HEX MEDIUM)

The Shigella detection medium is composed of proteose peptone (12.0 g), yeast extract (3.0 g), bile salts No. 3 (3.0 g), lactose (12.0 g), sucrose (12.0 g), D-xylose (12.0 g), salicin (2.0 g), sodium chloride (5.0 g), sodium thiosulfate (5.0 g), ferric ammonium citrate (1.5 g), agar (14.0 g), bromothymol blue (65.0 mg), acid fuchsin (0.1 g), and distilled water (1000 mL).

The HEX basal selective medium was prepared by heating with agitation till the medium was boiled. It was cooled to 50°C, and poured for culturing.[35] In the developing countries like India food contamination is very common, due to which, people often suffer from bacillary dysentery which is mainly caused by *S. flexneri.* Recently Wang et al. isolated a novel sRNA Ssr1 in *S. flexneri* using RT-PCR, northern blot, and 5′ RACE. Using murine lung invasion model and survival model analysis, they reported that the change in acidic environment can increase the virulence of Ssr1 mutant strain. Their study showed the environmental adaptation as well as pathogenic characteristics of a mutant strain of *S. flexneri* and shed light on its pathogenesis.[36] Figure 7.1 depicts the conventional cascade for *Shigella* detection in food samples.

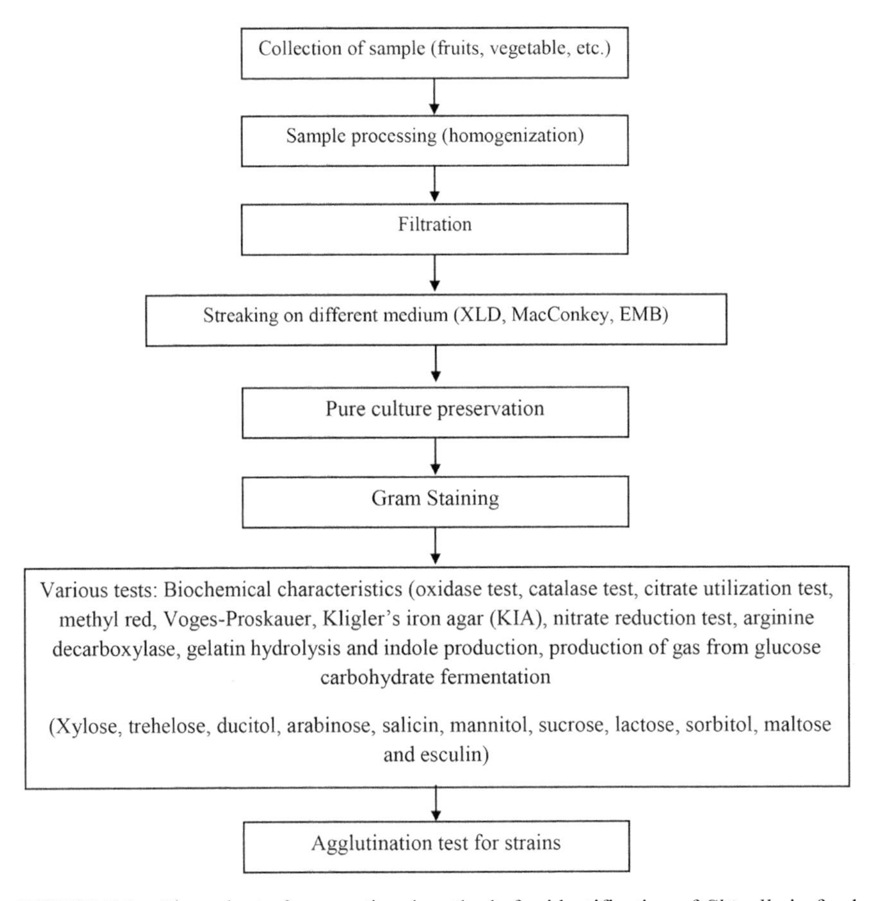

FIGURE 7.1 Flow chart of conventional methods for identification of *Shigella* in food sample.[33]

Nowadays, food safety is of great concern. To ensure the good quality of food, it is analyzed microbiologically using different conventional as well as novel rapid recent methods. Conventional methods are found to be laborious, tedious, and time consuming whereas, novel rapid recent methods are performed faster and efficiently to detect the pathogens in the food before it reaches the consumers. These new rapid recent methods include:

1. Molecular methods:
   a) Multiplex PCR[37]
   b) ATP bioluminescence[31]
2. Immunological methods:
   a) ELISA[31]
3. Biophysical and biochemical methods:
   a) Electrical impedometry[31]
   b) Flow cytometry[31]

7.4.1.1 MULTIPLEX PCR

Since past few years, multiplex PCR has been used for the detection of toxin producing foodborne pathogen[38] like *Listeria monocytogenes*, *Escherichia coli* O157:H7, *Staphylococcus aureus, Campylobacter jejuni, Salmonella* spp. and *Shigella* spp.[39–43] Through this method, several different DNA region of interest can be detected in a single PCR reaction using multiple primer pair. The conserved regions of specific genes for *S. flexneri, S. sonnei, S. boydii* such as *ipaH1, ipaH, wbgZ, and wzy* can easily be amplified with this method that helps in the detection of *Shigella* in food.[44] This molecular method is 100% efficient and accurate for *Shigella* detection in food.[37]

7.4.1.2 ATP BIOLUMINESCENCE

ATP bioluminescence is a technique by which ATP molecule of the microbial and non-microbial cells can be detected with the help of luminometers. To detect ATP of *Shigella*, the non-microbial ATP is first extracted using non-ionic detergents like potato ATPase for 5 min. Then the microbial ATP is extracted using either 5% trichloroacetic acid or organic solvent like ethanol, acetone, or chloroform. ATP molecule can be assayed using an enzyme coenzyme complex known as luciferase-luciferin. In total, 10^4 cells are required for a signal which is directly proportional to the ATP present in the cells. Here, 1 pg ATP is equivalent to 1000 cells.[31]

7.4.2 IMMUNOLOGICAL METHOD

The most sensitive method for the detection of foodborne pathogens is sandwich ELISA. In this method, using primary and secondary antibodies[45] the foodborne pathogens including bacteria like *Shigella* and its toxins can be identified. Horseradish peroxidase (HRP), alkaline phosphatase, and beta-galactosidase[46] are the enzymes used in ELISA method.

7.4.3 BIOPHYSICAL AND BIOCHEMICAL METHODS

7.4.3.1 ELECTRICAL IMPEDOMETRY

For food safety measures, detection of foodborne pathogens is also done using electrical impedance biosensor. This technique is simple and it can be measured pathogens rapidly. In this technique microbes like *Shigella* in food are analyzed in bactometer, which is computer driven and the measurement is done automatically in 24 h. To perform this process, the sample size should be high. In this method, the organisms are treated with non-electrolyte like lactose. Lactose produces lactic acid which changes the impedance and it can be monitored in a period of 20 h after inoculation has been done in specific media. Through this method hundreds of samples can be analyzed simultaneously at the same time.[31]

7.4.3.2 FLOW CYTOMETRY

The basic principle behind flow cytometry is based on scattering of light which can be detected by photomultiplier tubes using specific dye (fluid). Scattering of light is due to the presence of cells like *Shigella* in the medium that helps in detecting size, shape, and structure of microbes.[31]

7.4.3.3 PHAGE BIO-CONTROL

Phage bio-control technique is one of the recently applied techniques to maintain food safety. In this method, phages are used in livestock to reduce pathogenic contamination. Another application of bacteriophage is phage biosanitation. It is a surface sterilization method by which inanimate objects can be decontaminated from households, kitchen surfaces are made free of pathogenic bacteria like *Shigella*. During processing of ready-to-eat food, this method is used. Phages are also applied on the food after harvesting to save it from pathogenic attack. *Shigella* present in the processed meat is eliminated by using a single or mixture of phages.[47]

7.4.3.4 LOOP-MEDIATED ISOTHERMAL AMPLIFICATION (LAMP)

LAMP is a recent advanced approach for the detection of *Shigella* by targeting *ipaH* gene present in it. This method is an efficient one as the detection of gene is possible within 2 h in minimum number of bacteria like 8 cfu per reaction.[48]

Foodborne pathogens are considered as risk factor for food processing industries. To avoid loss because of contamination, many methods have been introduced to detect and prevent foodborne pathogens. Previously, conventional methods were applied in food industry. Although these methods maintain accuracy in detection they are time consuming and laborious. To avoid these disadvantages rapid methods were introduced that can detect pathogens in less time. Most of these methods are machine operated and are very sensitive, and accurate in detection of these pathogens.

7.5 PATHOGENESIS

The disease starts with internalization of the bacteria via endosome to attach with the target cell and enter the cytoplasm for multiplication. The organism incubates for 1–4 days. Colonization in colon causing damage in epithelial cells in colon which leads to form micro-ulcers, inflammatory exudates cause polymorphonuclear leukocytes (PMNs) in the lamina propria. Finally, the blood is appeared in stool which contain 10^6–10^8 organisms per gram. During the bloody diarrhea the volume of stool is dependent upon the ileocecal flow of colon resulting in scanty dysenteric stools.[2,49,50] The clinical manifestations include bacillary dysentery and invasive diarrhea including fever, abdominal pain, and tenesmus. In young children, toxic megacolon, bacteremia, Reiter's syndrome, and hemolytic-uremic syndrome are mostly reported.[51]

7.5.1 VIRULENCE FACTORS

7.5.1.1 CYTOTOXIN

Shiga toxin is the cytotoxin produced by *S. dysenteriae* serotype 1, *S. sonnei* and *S. flexneri*. The toxin produced by *S. dysenteriae* serotype 1 is more than *S. sonnei* and *S. flexneri*. The Shiga toxin inhibits protein synthesis by binding to the Galotl-4Galp (galabiose) glycolipid receptors and cleaves the N-glycosidic bond at adenine 4324 in 28S rRNA. The LPS endotoxin (O antigen) is released by the organism responsible for inflammation.[49,50]

7.5.1.2 PLASMID ENCODED GENES

The adherence and invasion of *Shigella* is encoded by plasmid which is 180–230 kb. Adhesions and invasion plasmid antigens (Ipa) are product of the plasmid which are responsible for its pathogenesis.[49,50]

7.6 FOOD ITEMS ASSOCIATED WITH *SHIGELLA* OUTBREAKS

Raw fruits and vegetables are important part of our diet because they are sources of valuable nutrients, micronutrients, vitamins, and fiber essential for good health and for reducing risk of deficiency diseases.[52] Consumption of fresh or minimally processed fruits and vegetables has increased due to awareness about the various health benefits associated with them. Restaurants and salad bars have become popular places where these food items are consumed away from home.[53] There is need for maintaining the quality and hygienic parameters of such food items to avoid risk of pathogenic bacteria like *Shigella* during consumption to obtain maximum health benefits. The main role for reducing risk of *Shigella* outbreaks lies with food producers, distributors, and vendors who need to follow safety norms during food production, transport, storage, and sale to maintain acceptable quality and safety.

A wide variety of fresh vegetables including leafy vegetables and tomato and berries has been implicated as sources of *Shigella* transmission and outbreaks in Canada. In multiple years of study it has been noted that summer season (July) has high incidence of outbreaks and detection of contamination with *Shigella.* Organic farming, conventional farming, use of animal manure, and fecal contamination were attributed for these outbreaks.[54] *Shigella* spp. is mostly encountered in fresh fruits and vegetables along with *E. coli* O157:H7, *Salmonella, L. monocytogenes*. The presence of *Shigella* spp. is a main food safety concern.

Several outbreaks involving *Shigella* Spp. were reported to be interprovincial and even international with same product as source of outbreak, including countries United States, Finland, and Denmark.[55] There are several pre-harvest and post-harvest activities and factors which lead to contamination and exposure of food items like fruits and vegetables to *Shigella* (Table 7.2).

TABLE 7.2 Source of *Shigella* Contamination and Exposure of Fruits and Vegetables.

Pre-harvest exposure and contamination	Post-harvest exposure and contamination
Soil	Handling at harvest
Air	Harvesting equipment
Dust	Transport containers
Irrigation water	Insect
Human and animal feces	Dust
Water used to apply fungicides and insecticides	Post-harvest washing
Insects	Rinse water quality
Inadequately composted manure	Ice
Organic manure	Improper cooking and/or improper holding temperature after cooking
Animal manure	Contamination from other foods in food preparation area
	Improper packaging
	Improper storage
	Cross contamination
	Hygiene of handler
	Transport vehicle
	Water use in processing
	Processing equipment
	Transportation
	Handling by individual retailers

Ready-to-eat fruit salads and vegetable salads have become very attractive due to the numerous health benefits and short preparation time. However, poor quality raw materials can harbor pathogenic microorganisms like *Shigella.* Lettuce, cucumber, tomato, spinach, cabbage, radish, etc., are used commonly during salad preparations. If the pH of fruits and vegetables is 4.6 or higher, then it is suitable for growth of pathogenic bacteria. There are also chances of cross contamination during preparation of mixed fruit or vegetable salads. Washing with water or disinfectants may be done to reduce microbial load. Pathogenic bacteria may infiltrate cracks, crevices of fruits and vegetables which make it hard to eliminate them. The biofilm present on surface of vegetables also provide a protective environment for pathogenic bacteria.

Moisture content, relative humidity, favorable pH, temperature, tissue damage, lesions, and nutrient-rich exudates support rapid multiplication of bacteria.[53] Several incidences of *Shigella* outbreaks all over the world are reported to be associated with vegetables and fruits such as lettuce, green onion, shredded cabbage, chopped parsley, scallion, jicama, cut papaya, and watermelon.[54,56] Lettuce and green pepper sampled from Addis Ababa, Ethiopia between November 2007and April 2008 were reported to contain multidrug resistant strains of *Shigella*. It was attributed to their use as raw salad vegetables without any heat treatment revealing the low quality and safety of these vegetables. The information on multidrug resistant pattern will be useful for antibiotic therapy against dysentery.[57] Studies on cucumber (*Cucumis satavum*), garden egg (*Salinum melongena*) carrot (*Daucus carota*), pumpkin (*Telfairia occidentalis*), waterleaf (*Talinum triangulare*), and afang (*Gnetum africanum*) from open area markets in Calabar Metropolis of Nigeria showed that *Shigella* spp. (57%) were found on the surface of waterleaf followed by pumpkin (33%).

According to USFDA, vegetables found associated with *Shigella* include cilantro, culantro, cantaloupe, tomatoes, loose-leaf lettuce (e.g., Romaine) or already separated leaves from the loose-leaf variety (e.g., mesclun, escarole, endive, witloof, chicory leaf, lamb's lettuce, mache, and radicchio), scallion (green onions), parsley, celery and broccoli. Contaminated lettuce led to an outbreak by *S. sonnei* at university campuses in the United States in 1985.[58] *S. sonnei* associated with iceberg lettuce from Spain was traced as the cause of infections in several European countries, including Norway, Sweden, and the United Kingdom in 1994.[59,60] In the same year, shigellosis outbreak in United States was linked to scallions (green onions) from United States and Mexico.[61]

In Europe, recent outbreaks have revealed new and unexplained links between *S.* sonnei and imported baby corn.[62] Minimally processed salads Tabbouleh, Fattoush, Hummus, Mutabbal, and Caesar which are consumed in Saudi Arabia were studied for contamination with *Shigella* and other foodborne pathogens. Ingredients of these salads were as follows: Tabbouleh—Parsley, bulgur, mint, tomato, scallion, herbs, lemon, black pepper, cinnamon, allspice, and lettuce; Fattoush—Radish, Tomato, Lettuce, Cucumber, Onion, Sumac, Mint, Olive oil, Lemon Juice, Salt, and Pita—toasted or fried; Hummus—Chickpeas, tahini, olive oil, lemon juice, salt, and garlic; Mutabbal—Eggplant, tahini, salt, pepper, olive oil, and pomegranate seeds; and Caesar—Romaine Lettuce, croutons, parmesan cheese, lemon juice, olive oil, egg, Worcestershire sauce, and black pepper. *Shigella* was detected in Tabbouleh, Hummus, and Mutabbel.[63]

Examination of ready to eat (RTE) vegetable salads prepared with beetroot, coriander, cucumber, carrot, fenugreek, radish, and spinach in Amravati, India, showed prevalence of *Shigella* spp. (3.4%) along with other enteric pathogens like *Enterobacter aerogenes*, *Pseudomonas* spp., *S. aureus* and *Salmonella* spp.[64] *Shigella* spp. grow in the pH range of 4.5–9, and can remain viable at pH 2–3 for several hours. Salads with acidic pH (commercial carrot salad pH 2.7–2.9, potato salad pH 3.3–4.4, coleslaw pH 4.1–4.2, and crab salad pH 4.4–4.5) can also be implicated in shigellosis outbreaks. *Shigella* can also survive at refrigeration temperatures under low O_2 packaging conditions. *Shigella* has also been reported to survive at 4°C in commercial salads for at least 11 days and up to 20 days in crab salad.[56] *S. flexneri* was implicated in fruit salad in the UK.[65] Rapid growth of *Shigella* occurs in shredded lettuce stored at warm temperatures. In salsa preparations, rapid growth of pathogen like *Salmonella* was observed if jalapeno peppers were added while slower growth was seen if fresh garlic and lime juice were added in the salsa mix.[66]

A study of the microbiological status of morning glory harvested from around Phnom Penh, Cambodia, and collected from vendors in the marketplace showed 15% of samples were contaminated with *Shigella*, 100% with *E. coli* and 29% with *Clostridium perfringens*.[62] *Shigella* spp. are able to survive lower pH conditions at reduced temperatures, with *S. flexneri* and *S. sonnei* viable for 14 days in tomato juice (pH 3.9–4.1) and apple juice (pH 3.3–3.4) stored at 7°C. *S. flexneri* is salt tolerant and is able to grow in media containing 7% NaCl at 28°C.[68]

Outbreaks of *S. sonnei* infection were associated with consumption of fresh chopped, uncooked curly parsley in United States and Canada. Chopped and uncooked parsley served on chicken sandwiches and in coleslaw was traced in Minnesota, United States as the source. In Ontario, Canada, smoked salmon and pasta dish made with fresh chopped parsley sold at a kiosk at a food fair was implicated as the source of outbreak. Changes in parsley production on the farm (e.g., the use of adequately chlorinated water for chilling and icing parsley, education of farm workers on proper hygiene, and possibly the use of post-harvest control measures such as irradiation) may be necessary to ensure that produce is not contaminated with pathogens.

Shigella spp. were found to be associated with beef carcasses in abattoirs of Zeerust and Rustenburg in the North West Province, South Africa. Such contamination of carcasses with intestinal content and excreta are risk factors for outbreaks of infection. *Shigella* species were detected in 73/97 (75.3%) of the samples analyzed. These findings were higher than those reported in other countries. In Egypt, 2.0% of meat products were positive for *Shigella* species while only a 0.6% prevalence rate was detected in meat products analyzed in Ethiopia.[69] *Shigella* isolates were obtained from chicken sold from open markets and cold stores where imported chicken is sold. Live birds and chicken carcasses were assessed and it was inferred that these also can be potential source of *Shigella,* among other enteropathogenic bacteria.[70,71] *Shigella* has also been a concern in raw and RTE meat.[72] After *E. coli* O157:H7 and *Salmonella, Shigella* has been implicated with outbreaks of foodborne infections by CDC.[73]

S. flexneri infection and outbreaks have been linked to unpasteurized apple juice and apple cider, tomatoes and tomato juice, and unpasteurized orange juice. Other pathogens also detected with *Shigella* include *E. coli* O157: H7, *L. monocytogenes* and *Salmonella. S. flexneri* and *S. sonnei* have been reported to survive for at least 14 days in apple juice and tomato juice (pH 3–4). The presence of pathogenic bacteria in soil, water, and air and contact and exposure of raw unprocessed fruits and vegetables with these elements lead to their contamination. When these fruits and vegetables are peeled and cut, the physical external barrier is removed against pathogenic bacteria.

S. flexneri, and *S. sonnei* have been reported in shrimps, mussels, and oysters in France and United States. Raw shellfish consumption is implicated in gastroenteritis cases. An outbreak of *S. sonnei* involving 24 persons who had consumed raw oysters in Texas, United States was traced to poor sanitary conditions of the boat from which the oysters were procured.[74]

7.7 TREATMENT AND PREVENTION

Rehydration of patients is the first important step in diarrhea of any etiology and helps to reduce the number of deaths due to diarrhea. The common method of treatment of shigellosis in children is administration of oral rehydration solutions to compensate for loss of fluid and electrolyte loss.[75] High protein diet especially in developing countries is recommended to prevent malnutrition and for better immunity.[76] The World Health Organization (WHO) recommends that all suspected cases of shigellosis based on clinical features should be treated with effective antimicrobials.[77] The type of antimicrobial drug used around the world for shigellosis varies as antibiotic resistance patterns have emerged. Some of the antibiotics and antimicrobials used to treat *S. dysentery* are: Beta-lactams: ampicillin, amoxicillin, third-generation cephalosporins (cefixime and ceftriaxone), and pivmecillinam (not available in the United States); quinolones: nalidixic acid, ciprofloxacin, norfloxacin, and ofloxacin; macrolides: azithromycin; and others: sulfonamides, tetracycline, cotrimoxazole, and furazolidone. There are many antimicrobial agents which are effective in the treatment of shigellosis. Despite this, there is also danger of an emerging drug resistance.[78,79] There are some antimicrobial agents which are not recommended as *Shigella* species are resistant to them. They include sulfonamides, tetracyclines, ampicillin, and trimethoprim-sulfamethoxazole (TMP-SMX).[78] WHO now recommends ceftriaxone, pivmecillinam and azithromycin as alternative drugs to fluroquinolone resistant shigellae.[80]

As a preventive measure, the public must be informed about how *Shigella,* and other organisms that cause diarrhea, are transmitted and how transmission can be prevented. In this regard, health education and public awareness are of prime importance. Hand-washing with soap after defecation or cleaning child after defecation, as well as before preparation of food should be promoted. Drinking water should be kept safe from fecal contamination during transport and storage. Disposal of feces should be done in sanitary systems appropriate for local conditions. The widespread practice of misuse of antibiotics in viral diarrhea should be discouraged by means of education as well as legislation.[80] Shigellosis is targeted by WHO as one of those enteric infections for which new vaccines are most needed, the target populations being travelers from developed countries and military service personnel, as well as children living in endemic areas.[12]

7.8 HACCP FOR *SHIGELLA* OUTBREAK MANAGEMENT

It is not practical to target bacterial pathogens like *Shigella* in cooking or pasteurization processes altogether. Such pathogens should be controlled through a rigorous sanitation regime or as part of HACCP. Food items that are cooked before packaged, are at risk of recontamination between cooking and packaging. The risk of recontamination may be minimized by filling the container in a sanitary, automated, and continuous filling system while the product is still hot. This may be a critical step for the safety of the packaged products. This control strategy is suitable for products that are filled directly from the cooking kettle, where the risk of recontamination is minimized.[81] The WHO, has stressed upon the importance and utility of HACCP for improving the safety of foods, has been encouraging the development and application of this system in the integrality of the food chain, from production until consumption.

KEYWORDS

- ***Shigella***
- **shigellosis**
- **Shiga toxin**
- **virulence**
- **food-borne illness**
- **HACCP**

REFERENCES

1. Identification of *Shigella* Species. UK Standards for Microbiology Investigations. Issued by the Standards Unit, Public Health England. Bacteriology-Identification. ID: 20. Issue no: 3. Available at: http://www.sfam.org.uk/download.cfm?docid=369E1393-62FA-46C1-B8BA070870C98F0B

2. Cabral, J. P. S. Water Microbiology. Bacterial Pathogens and Water. *Int. J. Environ. Res. Public Health.* **2010,** *7* (10), 3657–3703.
3. WHO (World Health Organization). *Guidelines for Drinking-water Quality, Incorporating 1st and 2nd Addenda, Recommendations,* 3rd ed.; WHO: Geneva, Switzerland, **2008**; Vol. 1.
4. Kapperud, G.; Rørvik, L. M.; Hasseltvedt, V.; Høiby, E. A.; Iversen, B. G.; Staveland, K.; Johnsen, G.; Leitão, J.; Herikstad, H.; Andersson, Y.; Langeland, G.; Gondrosen, B.; Lassen, J. Outbreak of *Shigella sonnei* Infection Traced to Imported Iceberg Lettuce. *J. Clin. Microbiol.* **1995,** *33,* 609–614.
5. Strockbine, N. A., Maurelli, A. T., Brenner, D. J., Krieg, N. R., Staley, J. T., Eds. Genus Shigella. In *Bergey's Manual of Systematic Bacteriology,* 2nd ed.; Part B; Brenner Springer: New York, NY, 2005; Vol. 2, pp 811–823.
6. Germani, Y.; Sansonetti, P. J. The Genus *Shigella.* The Prokaryotes: In *An Evolving Electronic Resource for the Microbiological Community, Electronic Release 3.14,* 3rd ed.; Dworkin, M., Falkow, S., Rosenberg, E., Eds.; Springer-Verlag: New York, NY, 2003.
7. Tetteh, G. L.; Beuchat, L. R. Survival, Growth, and Inactivation of Acid-stressed *Shigella flexneri* as Affected by pH and Temperature. *Int. J. Food Microbiol.* **2003,** *87,* 131–138.
8. Chompook, P.; Todd, J.; Wheeler, J. G.; Von Seidlein, L.; Clemens, J.; Chaicumpa, W. Risk Factors for Shigellosis in Thailand. *Int. J. Infect. Dis.* **2006,** *10,* 425–433.
9. Faruque, S. M.; Khan, R.; Kamruzzman, M.; Yamasaki, S.; Ahmad, Q. S.; Azim, T.; Nair, G. B.; Takeda, Y.; Sack, D. A. Isolation of *Shigella dysenteriae* Type 1 and *S. flexneri* Strains from Surface Waters in Bangladesh: Comparative Molecular Analysis of Environmental *Shigella* Isolates versus Clinical Strains. *Appl. Environ. Microbiol.* **2002,** *68,* 3908–3913.
10. Emch, M.; Ali, M.; Yunus, M. Risk Areas and Neighborhood-level Risk Factors for *Shigella dysenteriae* 1 and *Shigella flexneri. Health Place.* **2008,** *14,* 96–105.
11. Bennish, M. L.; Wojtyniak, B. J. Mortality Due to Shigellosis: Community and Hospital Data. *Rev. infect. dis.* **1991,** *13* (suppl. 4), S245–S251.
12. Kotloff, K. L.; Winickoff, J. P.; Ivanoff, B.; Clemens, J. D.; Swerdlow, D. L.; Sansonetti, P. J.; Adak, G. K.; Levine, M. M. Global Burden of *Shigella* Infections: Implications for Vaccine Development and Implementation of Control Strategies. *Bull World Health Organ.* **1999,** *77* (8), 651–666.
13. Nygren, B.; Bowen, A. Shigella. In *Food Infections and Intoxications*, 4th ed.; Morries, J. G., Jr. Potter, M. E., Eds.; Academic Press: London, 2013; pp 217–221.
14. Jung, L.; Ahn, J. Evaluation of Bacteriophage Amplification Assay for Rapid Detection of *Shigella boydii* in Food Systems. *Ann. Microbiol.* **2015,** *66,* 883–888. DOI: 10.1007/s13213-015-1178-y
15. DuPont, H. L.; Levine, M. M.; Hornick, R. B.; Formal, S. B. Inoculum Size in Shigellosis and Implications for Expected Mode of Transmission. *J. Infect. Dis.* **1989,** *159,* 1126–1128.
16. Gupta, A.; Polyak, C. S.; Bishop, R. D.; Sobel, J.; Mintz, E. D. Laboratory-confirmed Shigellosis in the United States, 1989–2002: Epidemiologic Trends and Patterns. *Clin. Infect. Dis.* **2004,** *38,* 1372–1377.

17. CDC (Centres for Disease Control and Prevention). *Shigella* Surveillance: Annual Summary (2002–2006). US Department of Health and Human Services: Atlanta, GA, 2008.
18. Garrett, V.; Bornschlegel, K.; Lange, D.; Reddy, V.; Kornstein, L.; Kornblum, J, et al. A Recurring Outbreak of *Shigella sonnei* among Traditionally Observant Jewish Chindren in New York City: The Risks of Daycare and Household Transmission. *Epidemiol. Infect.* **2006,** *134,* 1231–1236.
19. Gaudreau, C.; Ratnayake, R.; Pilon, P. A.; Gagon, S.; Roger, M.; Levesque, S. Ciprofloxacin-resistant *Shigella sonnei* among Men Who have Sex with Men, Canada, 2010. *Emerg. Infect. Dis.* **2011,** *17,* 1747–1750.
20. Nygren, B. L.; Schilling, K. A.; Blanton, E. M.; Silk, B. J.; Cole, D. J.; Mintz, E. D. Foodborne Out-breaks of Shigellosis in the United States, 1998–2008. *Epidemiol. Infect.* **2012,** *24,* 1–9.
21. Scallan, E.; Hoekstra, R. M.; Angulo, F. J.; Tauxe, R. V.; Widdowson, M. A.; Roy, S. L, et al. Foodbrone Illness Acquired in the United States-major Pathogens. *Emerg. Infect. Dis.* **2011,** *17,* 7–12.
22. Haley, C. C.; Ong, K. L.; Hedberg, K.; Cieslak, P. R.; Scallan, E.; Marcus, R, et al. Risk Factors for Sporadic Shigellosis, Food Net 2005. *Foodborne Pathog. Dis.* **2010,** *7,* 741–747.
23. Ghosh. S.; Pazhani, G. P.; Niyogi, S. K.; Nataro, J. P.; Ramamurthy, T. Genetic Characterization of *Shigella* spp. Isolated from Diarrhoeal and Asymptomatic Children. *J. Med. Microbiol.* **2014,** *63,* 903–910.
24. Ke, X.; Gu, B.; Pan, S.; Tong, M. Epidemiology and Molecular Mechanism of Integron-mediated Antibiotic Resistance in *Shigella. Arch. Microbiol.* **2011,** *193,* 767–774.
25. Robicsek, A.; Strahilevitz, J.; Jacoby, G. A.; Macielag, M.; Abbanat, D.; Park, C. H.; Bush, K.; Hooper, D. C. Fluoroquinolone-modifying Enzyme: A New Adaptation of a Common Aminoglycoside Acetyltransferase. *Natr. Med.* **2006,** *12* (1), 83–88.
26. Pu, X. Y.; Pan, J. C.; Wang, H. Q.; Zhang, W.; Huang, Z. C.; Gu, Y. M. Characterization of Fluoroquinolone-resistant *Shigella flexneri* in Hangzhou Area of China. *J. Antimicrob. Chemother.* **2009,** *63,* 917–920.
27. Bhattacharya, D.; Bhattacharjee, H.; Thamizhmani, R.; Sayi, D. S.; Bharadwaj, A. P.; Singhania, M, et al. Prevalence of the Plasmid Mediated Quinolone Resistance Determinants among Clinical Isolates of *Shigella* sp. in Andaman & Nicobar Islands, India. *Lett. Appl. Microbiol.* **2011,** *53,* 247–251.
28. Vunrcrzant, C.; Pllustoesser, D. F. *Compendium of Methods for Microbial Examination of Food,* 3rd ed.; American Public Health Association: New York, NY, 1987.
29. Davidson, P. M.; Doyle, M.; Beuchat, L.; Montville, T. Food Microbiology-fundamentals and Frontiers. *Chem. Pres. Natr. Antimicrobial Cpd.* **2001**, 593–627.
30. Ivnitski, D.; Abdel-Hamid, I.; Atanasov, P.; Wilkins, E. Biosensors for Detection of Pathogenic Bacteria. *Biosens. Bioelectron.* **1999**, *14* (7), 599–624.
31. Biswas, A. K. Evaluation of Export Buffalo Meat for Some Chemical Residues and Microbial Contaminants. (Doctoral dissertation, Indian Veterinary Research Institute; Izatnagar, Bareilly, Uttar Pradesh, India, 2005.
32. Mandal, P. K.; Biswas, A. K.; Choi, K.; Pal, U. K. Methods for Rapid Detection of Foodborne Pathogens: An Overview. *Am. J. Food Technol.* **2011**, *6* (2), 87–102.

33. Gundappa, M.; Gaddad, S. M. Prevalence of Salmonella, Shigella and E. coli in Vegetables of Various Markets in Kalaburagi (India). *Indian J. Nat. Sci.* **2016,** *6* (1), 1–5.
34. Fakruddin, M.; Rahaman, M. M.; Ahmed, M. M.; Hoque, M. M. Occurrence of Enterobacteriaceae with Serological Cross Reactivity towards *Salmonella* spp., *Shigella* spp. and *Vibrio cholerae* in Food. *Brit. Microbiol. Res. J.* **2015,** *5* (1), 44–51.
35. Lee, Ji-Y.; Kand, D. H. Development of an Improved Selective Medium for the Detection of *Shigella spp. LWT Food Sci. Technol.* **2016,** *62*, 311–317.
36. Wang, L.; Yang, G.; Qi, L.; Li, X.; Jia, L.; Xie, J.; Qiu, S.; Li, P.; Hao, R. J.; Wu, Z.; Du, X.; Li, W.; Song, H. A Novel Small RNA Regulates Tolerance and Virulence in *Shigella flexneri* by Responding to Acidic Environmental Changes. *Front. Cell. Infect. Microbiol.* **2016,** *6* (24), 1–12.
37. Jimenez, K. B.; McCoy, C. B.; Achi, R. Detection of *Shigella* in Lettuce by the Use of a Rapid Molecular Assay with Increased Sensitivity. *Braz. J. Microbiol.* **2010,** *41,* 993–1000.
38. Law, J. W. F.; AbMutalib, N. S.; Chan Gand, K.; Lee, L. H. Rapid Methods for the Detection of Food Borne Bacterial Pathogens: Principles, Applications, Advantages and Limitations. *Front. Microbiol.* **2014,** *5,* 770.
39. Cheah,Y. K.; Noorzaleha, A. S.; Lee, L. H.; Radu, S.; Sukardi, S.; Sim, J. H. Comparison of PCR Finger Printing Techniques for the Discrimination of *Salmonella enterica* subsp. *enterica* Serovar Weltevreden Isolated from Indigenous Vegetables in Malaysia. *World J. Microbiol. Biotechnol.* **2008,** *24,* 327–335. doi: 10.1007/s11274-007-947
40. Lee, L. H.; Cheah, Y. K.; Noorzaleha, A. S.; Sabrina, S.; Sim, J. H.; Khoo, C. H, et al. Analysis of *Salmonella* Agona and *Salmonella* Weltevredenin Malaysia by PCR Fingerprinting and Antibiotic Resistance Profiling. *Antonie Van Leeuwenhoek.* **2008,** *94,* 377–387. doi:10.1007/s10482-008-9254-y
41. Alves, J.; Marques, V. V.; Pereira, L. F. P.; Hirooka, E. Y.; Moreirade Oliveira, T. C. R. Multiplex PCR for the Detection of *Campylobacter spp.* and *Salmonella spp.* In Chicken Meat. *J. Food Saf.* **2012,** *32,* 345–350. doi:10.1111/j.1745-4565.2012.00386.x
42. Chiang, Y. C.; Tsen, H. Y.; Chen, H. Y.; Chang, Y. H.; Lin, C. K.; Chen, C. Y.; Pai, W. Y. Multiplex PCR and Achromogenic DNA Macro Array for the Detection of *Listeria monocytogenes, Staphylococcus aureus, Streptococcus agalac-tiae, Enterobacter sakazakii, Escherichia coli* O157:H7, *Vibrio parahaemolyticus*, *Salmonella spp.* and *Pseudomonas fluorescens* in Milk and Meat Samples. *J. Microbiol. Methods.* **2012,** *88,* 110–116.
43. Zhou, B.; Xiao, J.; Liu, S.; Yang, J.; Wang, Y.; Nie, F. Simultaneous Detection of Six Food-borne Pathogens by Multiplex PCR with GEXP Analyzer. *Food Control.* **2013,** *32,* 198–204.
44. Radhika, M.; Saugata, M.; Murali, H. S.; Batra, H. V.; A Novel Multiplex PCR for the Simultaneous Detection of *Salmonella enteric* and *Shigella* Species. *Braz. J. Microbiol.* **2014,** *45* (2), 667–676.
45. Zhao, X.; Lin, C. W.; Wang, J.; Oh, D. H. Advances in Rapid Detection Methods for Foodborne Pathogens. *J. Microbiol. Biotechnol.* **2014,** *24* (3), 297–312.
46. Yeni, F.; Acar, S.; Polat, O. G.; Soyer, Y.; Alpas, H. Rapid and Standardized Methods for Detection of Food Borne Pathogens and Mycotoxins on Fresh Produce. *Food Control.* **2014,** *40,* 359–367.

47. Sulakvelidze, A. Using Lytic Bacteriophages to Eliminate or Significantly Reduce Contamination of Food by Foodborne Bacterial Pathogens. *J. Sci. Food Agric.* **2013,** *93* (13), 3137–3146.
48. Song, T.; Toma, C.; Nakasone, N.; Iwanaga, M. Sensitive and Rapid Detection of *Shigella* and Enteroinvasive *Escherichia coli* by a Loop-mediated Isothermal Amplification. 2005. *FEMS Microbiol. Lett.* **2005,** *243* (1), 259–263.
49. Hale, T. L. Genetic Basis of Virulence in *Shigella* Species. *Microbiol. Rev.* **1991,** *55,* 206–224.
50. Todar, K. Shigella and Shigellosis. *Todar's Online Textbook of Bacteriology* [Online]; 2009. http://www.textbookofbacteriology.net/Shigella.html. (assessed on Sep 1, 2010).
51. Iwamoto, M.; Ayers, T.; Mahon, B. E.; Swerdlow, D. L.; Epidemiology of Seafood-associated Infections in the United States. *Clin. Microbiol. Rev.* **2010,** *23* (2), 399–411.
52. Obi, C. N. Bacteriological Assessment of Vegetables Cultivated in Soils Treated with Poultry Manure and the Manure-treated Soil Samples. *Am. J. Microbiol. Res.* **2014,** *2* (6), 189–200.
53. Buck, J. W.; Walcott, R. R.; Beuchat, L. R. Recent Trends in Microbiological Safety of Fruits and Vegetables. *Plant. H. Prog.* **2003**, *10* (1), 1094.
54. Denis, N.; Zhang, H.; Leroux, A.; Trudel, R.; Bietlot, H. Prevalence and Trends of Bacterial Contamination in Fresh Fruits and Vegetables Sold at Retail in Canada. *Food Control.* **2016,** *67,* 225–234.
55. Sewell, A. M.; Farber, J. M.; Food Borne Outbreaks in Canada Linked to Produce. *J. Food Protect.* **2001,** *64* (11), 1863–1877.
56. Selmaa, M. V.; Beltrana, D.; Allendea, A.; Verab, E. C.; Gila, M. I. Elimination by Ozone of *Shigella sonnei* in Shredded Lettuce and Water. *Food Microbiol.* **2007,** *24,* 492–499.
57. Guchi, B.; Ashenafi, M. Microbial Load, Prevalence and Anti Biograms of *Salmonella* and *Shigella* in Lettuce and Green Peppers. *Ethiop. J. Health Sci.* ***2010,*** *20* (1), 41–48.
58. Martin, D. L.; Gustafson, T. L.; Pelosi, J. W.; Suarez, L.; Pierce, G. V. Contaminated Produce--a Common Source of Two Outbreaks of *Shigella gastroenteritis*. *Am. J. Epidemiol.* **1986,** *124,* 299–305.
59. Frost, J. A.; McEvoy, M. B.; Bentley, C. A.; Andersson, Y.; Rowe, B. An Outbreak of *Shigella sonnei* Infection Associated with Consumption of Iceberg Lettuce. *Emerg. Infect. Dis.* **1995,** *1,* 26–29.
60. Kapperud, G.; Rorvik, L. M.; Hasseltvedt, V.; Hoiby, V.; Iverson, B. G.; Staveland, K.; Johnsen, G.; Leitao, J.; Herikstad, H.; Andersson, Y.; Langeland, G.; Gondrosen, B.; Lassen. J. Outbreak of Shigella Sonnei Infection Traced to Imported Iceberg Lettuce. *J. Clin. Microbiol.* **1995,** *33,* 609–614.
61. Cook, K. A.; Boyce, T.; Langkop, C.; Kuo, K.; Swartz, M.; Ewert, D.; Sowers, E.; Wells, J.; Tauxe, R. Scallions and Shigellosis: *A Multistate Outbreak Traced to Imported Green Onions. 36,* Epidemic Intelligence Service 44th Annual Conference, Mar 27–31, 1995; CDC: Atlanta, GA, 1995.
62. Lynch, M. F.; Tauxe, R. V.; Hedberg, C. W. The Growing Burden of Foodborne Outbreaks Due to Contaminated Fresh Produce: Risks and Opportunities *Epidemiol. Infect.* **2009,** *137,* 307–315.

63. Khiyami, M.; AL-Faris, N.; Busaeed, B.; Sher, H. Food Borne Pathogen Contamination in Minimally Processed Vegetable Salads in Riyadh, Saudi Arabia. *J. Med. Plants Res.* **2011,** *5* (3), 444–451.
64. Tambekar, D. H.; Mundhada, R. H. Bacteriological Quality of Salad Vegetables Sold in Amravati City (India). *J. Biol. Sci.* **2006,** *6* (1), 28–30.
65. Beuchat, L. R. Ecological Factors Influencing Survival and Growth of Human Pathogens on Raw Fruits and Vegetables. *Microbes Infect.* **2002,** *4,* 413–423.
66. Berger, C. N.; Sodha, S. V.; Shaw, R. K.; Griffin, P. M.; Pink, D.; Hand, P.; Frankel, G. Fresh Fruit and Vegetables as Vehicles for the Transmission of Human Pathogens. *Environ. Microbiol.* **2010,** *12* (9), 2385–2397.
67. WHO Regional Office for the Western Pacific. Final Report on the Microbiological Status of Morning Glory in Cambodia. Prepared by Dr Pau Ann Sivutha, Food Safety Bureau, Department of Drugs and Food, Ministry of Health Cambodia, 2005.
68. Girma, G. Prevalence, Antibiogram and Growth Potential of *Salmonella* and *Shigella* in Ethiopia: Implications for Public Health: A Review. *Res. J. Microbiol.* **2015,** *10* (7), 288–307.
69. Makabanyane, I. N.; Ndou, R. V.; Ateba, C. N. Genotypic Characterization of Shigella Species Isolated from Abattoirs in the North West Province, South Africa Using PCR Analysis. *J. Food Nutr. Res.* **2015,** *3* (2), 121–125.
70. Heredia, N.; Garcia, S.; Rojas, G.; Salazar, L. Microbiological Condition of Ground Meat Retailed in Monterrey, Mexico. *J. Food Protect.* **2001,** *64,* 1249–1251.
71. Sackey, B. A.; Mensah, P.; Collison, E.; Dawson, E. S. *Campylobacter*, *Salmonella*, *Shigella* and *Escherichia coli* in Live and Dressed Poultry from Metropolitan Accra. *Int. J. Food Microbiol.* **2001,** *71,* 21–28.
72. Acheson, D. W. K. Shigella. In *Guide to Foodborne Pathogens;* Labbe, R., Garcia, S., Eds.; Wiley: New York, NY, 2001; pp 193–200.
73. CDC. U.S. Foodborne Disease Outbreaks, Searchable Data 1990–1995. [Online]; 2002. Available from (http://www2a.cdc.gov/ncidod/foodborne/OutbreaksReport.asp)
74. Reeve, G.; Martin, D. L.; Pappas, J.; Thompson, R. E.; Greene, K. D. An Outbreak of Shigellosis Associated with the Consumption of Raw Oysters. *N. Eng. J. Med.* **1990,** *321,* 224–227.
75. Edwards, B. H. *Salmonella* and *Shigella* Species. *Clin. Lab. Med.* **1999,** *19,* 469–487.
76. Kabir, I.; Rahman, M. M.; Haider, R. Increased Height Gain of Children Fed a High-protein Diet during Convalescence from Shigellosis: A Six Month Follow-up Study. *J. Nutr.* **1998,** *128,* 1688–1691.
77. Christopher, P. R.; David, K. V.; John, S. M.; Sankarapandian, V. Antibiotic Therapy for *Shigella* Dysentery. *Cochrane Database Syst. Rev.* **2010,** *4,* CD006784.
78. Ashkenazi, S.; Levy, I.; Kazaronovski, V. Growing Antimicrobial Resistance of *Shigella* Isolates. *J. Antimicrob. Chemother.* **2003,** *51,* 427–429.
79. Bhattacharya, S. K.; Sur, D. An Evaluation of Current Shigellosis Treatment. *Expert Opin. Pharmacother.* **2003,** *4,* 1315–1320.
80. World Health Organization. The Treatment of Diarrhea: A Manual for Physicians and Other Senior Health Workers, 4th Revision. Geneva, **2005**, ISBN 94 4 159318 0.
81. Frazier, J. *Establishing or Verifying a Heat Process for Cooked, Ready-to-eat Seafood Products, and Heat Process Monitoring Considerations Under HACCP,* 2nd ed.; Grocery Manufacturers Association (Food Products Association): Washington, DC, 2005.

PART III
Foodborne Viruses

CHAPTER 8

NOROVIRUS

YUAN HU[1*] and HAIFENG CHEN[2*]

[1]*Northeast Region Laboratory, Office of Regulatory Affairs, Food and Drug Administration, Jamaica, NY, USA*

[2]*Division of Molecular Biology, Office of Applied Research and Safety Assessment, Center for Food Safety and Applied Nutrition, Food and Drug Administration, Laurel, MD, USA*

[*]*Corresponding author. E-mail: yuan.hu@fda.hhs.gov; haifeng.chen@fda.hhs.gov*

CONTENTS

Abstract 206
8.1 Introduction 206
8.2 Biological Characteristics 207
8.3 Pathogenesis 210
8.4 Norovirus Illness 210
8.5 Epidemiology 211
8.6 Laboratory Testing 212
8.7 Treatment 218
8.8 Prevention 218
8.9 Lessons Learned from Norovirus Outbreaks and Surveillance Information 220
Acknowledgments 232
Keywords 232
References 232

ABSTRACT

Human norovirus is a major cause of foodborne acute gastroenteritis and poses a significant public health concern worldwide. The detection and identification of norovirus in clinical specimens, food, and environmental samples have been mainly determined by molecular methods. Real-time quantitative polymerase chain reaction (RT-qPCR) has become a method of choice for molecular detection of norovirus in terms of sensitivity, specificity, and assay turnaround time. The rapidly emerging application of whole-genome sequencing (WGS) provides maximum resolution for nucleic acid-based characterization of norovirus strains, which will tremendously improve our ability to detect foodborne disease outbreaks. In the absence of licensed vaccines, the ways to control and prevent norovirus infection still relies on good person and food hygiene, good industrial/agricultural practice, effective product management, and development of improved monitoring techniques for detecting the virus.

8.1 INTRODUCTION

The "winter vomiting disease" was first described in 1929,[1] but the causative viral agent has not been identified until 1972.[2] The responsible viral pathogen is named norovirus (formerly known as Norwalk-like virus) that is a single-stranded, non-enveloped RNA virus belonging to the *Caliciviridae* family. Human norovirus, an enteric pathogen, is transmitted primarily through fecal-oral route either by direct contact of infected individuals or by consumption of contaminated food and water, making it an important foodborne viral pathogen. Norovirus is the most common cause of acute nonbacterial gastroenteritis in the United States, estimated to be responsible for nearly 60% of foodborne illnesses.[3] Norovirus has become a leading cause of both epidemic and sporadic acute gastroenteritis and is responsible for at least 50% of all gastroenteritis outbreaks in the world.[3] Each year, it causes 19–21 million illnesses and contributes to 56,000–71,000 hospitalizations as well as 570–800 deaths.[3] In spite of significant economic burden and morbidity caused by the virus, there are currently no antiviral agents or vaccines that can be administered for treatment or prevention of norovirus infections. In most cases, the only treatment required is dehydration therapy. Due to the lack of robust cell culture systems or small animal models to cultivate the virus, many aspects of norovirus biology are still not well studied and characterized. However, the recent application of genomic technologies has led to an increased understanding of the virus molecular biology, and this has resulted in the development and use of sensitive molecular methods for detection and identification of norovirus.[4,5] This chapter will provide a brief summary of methods for detecting, controlling, and preventing human norovirus infection as well was regulatory surveillance information on norovirus outbreaks.

8.2 BIOLOGICAL CHARACTERISTICS

Human norovirus is the most common cause of non-bacterial foodborne-disease outbreaks in the world. However, since the discovery of the virus in 1972, human norovirus has not been able to replicate in cell culture. The inability to replicate human norovirus has resulted in the use of potential norovirus surrogates which include murine norovirus (MNV), feline calicivirus (FCV), and Tulane virus, all of which can be replicated in cell culture. MNV shares many biological and molecular characteristics with human norovirus and thus has become the model of choice for the majority of molecular studies.[6,7] The MNV model system offers the first opportunity to understand the basic mechanisms of norovirus biology and pathogenesis in a natural host.[8,9] Another useful tool for studying virus-host interactions is the use of recombinant virus-like particles (VLPs), which were first generated when the capsid gene of norovirus was expressed in baculovirus system.[10] VLPs are structurally and antigenically similar to their corresponding wild-type viruses, making them serve as an important reagent for immunology studies or the development of diagnostic assays as well as candidate vaccines.

8.2.1 GENOME

Norovirus is a single-stranded, linear, and positive sense RNA virus. The total genome is ~7600 nucleotides in length. The 5' terminus has a genome-linked protein (VPg) and the 3' terminus has a poly (A) tract. Norovirus genome contains three open reading frames (ORFs). The ORF1 (~5 kb) encodes the nonstructural polyprotein that is cleaved by viral 3C-like protease into probably six proteins, including the deduced RNA-dependent RNA polymerase. ORF2 (~1.8 kb) and ORF3 (~0.6 kb) encode the major (VP1) and minor (VP2) capsid proteins, respectively.[11]

8.2.2 CLASSIFICATION

Norovirus belongs to genus *Norovirus* in the family *Caliciviridae*, which includes a group of viruses that are about 27–40 nm in diameter and contain a 60 kDa single major capsid protein that carries 32 shallow, cup-like circular indentations.[11] The family comprises at least five genera including (1) Vesivirus that includes FCV, vesicular exanthema of swine virus, and San Miguel sea lion virus; (2) Lagovirus that includes rabbit hemorrhagic disease virus and European Brown hare syndrome virus; (3) Norovirus, a cause of gastroenteritis in humans; (4) Sapovirus that primarily causes diarrheic infections in humans and some other animal species; and (5) Nebovirus that is associated with enteric disease in cattle.[12]

8.2.2.1 PROTOTYPE

In 1968, an outbreak of acute gastroenteritis was reported in an elementary school in Norwalk, OH.[13] Kapikian et al. identified the 27 nm viral particles, named Norwalk virus, to be responsible for the outbreak of gastroenteritis.[2] The Norwalk virus was the first virus that was identified as causing human acute gastroenteritis. Although other small and round-structured viruses morphologically similar to Norwalk virus have been identified to be related to outbreaks of acute gastroenteritis, Norwalk virus was still considered the prototype of such viral agents.[14–16] These viral particles were described as 27 nm in their shortest diameter and 32 nm in their longest diameter (Fig. 8.1).

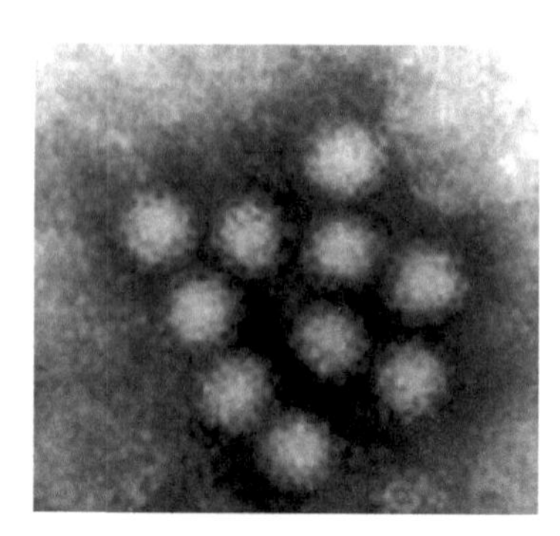

FIGURE 8.1 Electron micrograph of classical norovirus particles. (Source: F.P. Williams, U.S. Environmental Protection Agency)

8.2.2.2 GENOGROUPS

Norovirus is a group of highly related and very contagious viruses that can infect age group people. The diversity among norovirus strains is great and they can genetically be classified into at least five different genogroups, designated GI, GII, GIII, GIV, and GV, based on the complete amino acid identity in the major structural capsid protein (VP1).[17] Virus variants consisting of canine norovirus were recently discovered.[18,19] The viruses that infect humans (referred to collectively as "human noroviruses") are found in GI, GII, and GIV, whereas viruses in other genogroups are associated with infection of animals including rattle, mice, pigs, and dogs.

8.2.2.3 GENOTYPES

Noroviruses are highly diverse, which can be classified further into different genetic clusters or genotypes, with at least nine genotypes in GI and 22 genotypes in GII.[20] Despite the diversity, norovirus strains belonging to the GII are found in 75–100% of norovirus infection cases, and GII/4 variants have become the predominant causative agent for the most outbreaks worldwide since the mid-1990s (Fig. 8.2).[20,21] The novel GII.17 has been recently reported as dominant outbreak genotype in China,[22] replacing the previous pandemic GII.4 Sydney 2012. These emergent GII.17 strains have also been found in several countries and regions including the United States and Europe.[23,24] It is important to classify noroviruses correctly due to the emergence of new pandemic variants.

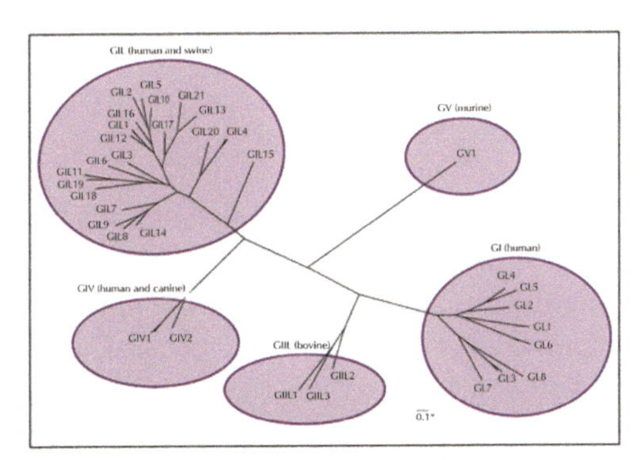

FIGURE 8.2 Classification of noroviruses into five genogroups (GI–GV) and 35 genotypes. (Source: Centers for Disease Control and Prevention 2014)

8.3 PATHOGENESIS

Noroviruses belonging to GI, GII, and GIV infect humans and cause acute gastroenteritis, while GIII and GV are found in cattle and mice, respectively. GII viruses also infect pigs. Canine norovirus is a pathogen recently discovered in dogs.[19] When norovirus infects humans, the virus interacts with the human histo-blood group antigens (HBGAs) as receptors or ligands for attachment, an early infection event that most likely controls host susceptibility and resistance to norovirus.[25] HBGAs are polymorphic and complex carbohydrates on red blood cell surfaces, saliva, and mucosal epithelia, which are linked to the ABO, secretor, and Lewis families. The interactions between noroviruses and HBGAs have been demonstrated to be essential for certain norovirus strains.[25] Different viral genotypes have different affinity for ABO antigens. GI noroviruses preferentially recognize blood group antigens A and O, while GII noroviruses prefer to bind blood group antigens A and B.[25,26] The binding domains for these antigens are located in the P2 region of the viral capsid. Blood group antigens likely serve as receptors for noroviruses or some other function critical for infection because human blood group type is closely linked to susceptibility to norovirus gastroenteritis. Norovirus infect cells via receptor-mediated attachment and clathrin-dependent endocytosis. Replication then occurs on the surface of membranous vesicles in the cytoplasm of infected cells.[11] Upon infection, the norovirus attaches to the outside cell lining of the intestine, and then transfers its genetic materials into those cells. The exact viral mechanisms which cause vomiting and diarrhea are unknown. The virus replication may directly kill the infected cells or the anti-noroviruses immune response destroys them.

8.4 NOROVIRUS ILLNESS

Norovirus causes acute gastroenteritis in persons of all ages. Onsets of symptoms including vomiting and diarrhea appear within 24 h of viral infection, and remain through the height of the illness, persisting for a variable period of time after the illness. In general, norovirus gastroenteritis is mild and self-limited, and most that contract it can make a full recovery within 1–3 days.[3] However, it may be more severe and prolonged in young children under five, elderly, and individuals with medical comorbidities since they are more liable to complications due to dehydration.[27]

Norovirus is a very contagious. The infectious dose can be as low as approximate 17 virus particles.[28] The incubation period lasts between 6 and 48 h. After that, the viruses infect stomach, intestines, or both to create acute gastroenteritis, which leads to abdominal pain, nausea, vomiting, and intense non-bloody diarrhea which generally resolves in 2–3 days. Diarrhea can lead to dehydration, especially in young children, elderly, and immunosuppressed individual. The patients are most contagious during the active disease phase and during the first few days after they recover from norovirus illness. Prolonged shedding is reported in immunocompromised patients; with symptoms lasting more than two years.[29] The virus can survive for long periods outside a human host depending on the surface and temperature conditions. It can stay in infected individuals' stool for 2 weeks or longer. Repeated infection may occur due to the lack of comprehensive cross-protection against the diverse norovirus strains and lasting immunity.[30] Being infected with one type of norovirus may not protect the patient against other types of viruses. Therefore, people may contract norovirus illness multiple times throughout life with re-exposure.

8.5 EPIDEMIOLOGY

Norovirus is the leading cause of foodborne viral illness and outbreaks in the United States.[3] Transmission of norovirus occurs through the fecal-oral route either by person-to-person contact or via consumption of contaminated food, water or exposure to airborne vomitus droplets, and contaminated environmental surfaces.[4] As of 2011, 20 states and local health laboratories had been certified to submit norovirus sequences and epidemiologic outbreak data to outbreak surveillance network.[31] On average, 365 foodborne norovirus outbreaks were reported annually to Centers for Disease Control and Prevention (CDC) Foodborne Disease Outbreak Surveillance System between 2001 and 2008.[3] Norovirus outbreaks involve people of all ages. Guidance for norovirus outbreak management and disease prevention in non-healthcare settings can be found at http://www.cdc.gov/mmwr/pdf/rr/rr6003.pdf. Other organizations that collaborate with CDC on outbreak surveillance and prevention include the Council of State and Territorial Epidemiologists (CSTE), the United States Environmental Protection Agency (EPA), the United States Department of Agriculture (USDA), and the United States Food and Drug Administration (FDA).

8.5.1 PERSON-TO-PERSON CONTACT

The majority of the outbreaks of norovirus illness occur when infected people spread the virus to others. Person-to-person contact is the most common route of norovirus transmission, accounting for 83.7% of all reported outbreaks in the United States from 2009 to 2013.[32] GII.4 appeared to be more frequently associated with person-to-person spread.[32] Norovirus outbreaks often occur in semi-close community settings, for example, schools, day-care centers, cruise ships and restaurants, and healthcare settings, for example, nursing homes and hospitals that favor person-to-person transmission. Most norovirus outbreaks happen from November to April in the United States.

8.5.2 NOROVIRUS AND FOOD CONTAMINATION

Foodborne transmission is an important route for the spread of noroviruses. In general, norovirus is persistent in the environment and can survive for a long period of time in water, food, and soil which facilitate viral transmission, making it an important foodborne and waterborne pathogen worldwide. Globally, approximately 14% of all norovirus outbreaks are attributed to food.[33] Food contamination with norovirus can be at any stage when the food is being grown, shipped, handled, or processed. For example, food produce is irrigated with contaminated water; food is processed by infected pickers or preparers; or contamination is as the consequence of adulteration during any point of handling.[34,35] Most norovirus outbreaks occur in the food service settings where the food is handled and prepared by infected food workers. Foods which are served raw or undercooked, such as molluscan shellfish, vegetable, or fruit salads have been frequently implicated in norovirus outbreaks.

8.6 LABORATORY TESTING

Initially, electron microscopy was the classic diagnostic method used to directly detect virus particles with 27–30 nm in diameter, the so-called small round structured viruses. As diagnostic technologies evolve, enzyme immunoassays (EIAs) have been made commercially available to detect norovirus GI and GII antigens from feces. The inability to cultivate human norovirus has resulted in the use of more sensitive molecular methods for detection from clinical, food, and environmental samples. Detection methods have improved significantly over the past two decades as a result of numerous molecular biology technological advances.

8.6.1 SAMPLE PREPARATION

Sample preparation has a vital role in overall detection and analysis. Norovirus can be detected in patient samples including stool, serum, and vomitus. Stool samples are the preferred clinical specimen due to the high viral loads in them. Many norovirus outbreaks have been linked to fresh and frozen produce which can become contaminated with norovirus either during handling and preparation, often by infected food handlers, or at source in the growing and harvesting area.[34,35] Norovirus is generally found in low numbers in contaminated food since the virus is not able to replicate without a living host. Compared to clinical samples, detection of norovirus in food is often compromised by low viral recovery and the presence of inhibitory materials. In general, detection of norovirus contamination in food requires separation, concentration, and purification of the virus from food matrices prior to molecular analysis. In this step, the virus is isolated from the food matrices and concentrated in a small final volume. Following the separation and concentration, viral RNA is then purified from the virus and subject to downstream molecular analysis. Currently, three major concentration methods have been used to recover and concentrate the viruses from food: (1) Polyethylene glycol (PEG) plus NaCl; (2) Ultrafiltration; and (3) Ultracentrifugation.[36,37]

8.6.2 ELECTRON MICROSCOPY

Norovirus was first identified in the 1970s using immune electron microscopy, which involved the direct observation of antigen–antibody complexes.[2] Electron microscopy was used for the diagnosis of norovirus infections and allowed direct visualization of the virus particles. However, this method lacks sufficient sensitivity and requires expensive instruments and well-trained microscopist, making clinical, or epidemiological studies impractical.[38]

8.6.3 ENZYME IMMUNOASSAYS

EIAs, first developed during the late 1970s, improved virus detection rates compared to electron microscopy.[3] EIAs have been recently made commercially available to detect norovirus GI and GII antigens from feces. They include the IDEIA NLV kit from Dako Cytomation, Ltd. (Ely, UK), the SRSV(II)-AD kit from Denka Seiken Co., Ltd. (Tokyo, Japan), and the RIDASCREEN norovirus kit from R-Biopharm AG (Darmstadt, Germany). The RIDASCREEN, a norovirus 3rd generation commercially available EIA kit, was cleared by FDA for preliminary screening of norovirus for testing multiple specimens during outbreaks. In general, these kits have low diagnostic sensitivity and are not recommended for diagnosing norovirus infection in sporadic gastroenteritis cases.[3] Samples tested negative should be confirmed by a second technique, such as real-time quantitative polymerase chain reaction (PCR) (RT-qPCR) due to their low diagnostic sensitivity.[3,39]

8.6.4 MOLECULAR TESTS

The breakthrough cloning and sequencing of the prototype strain Norwalk virus in 1990 led to the development of molecular tools to study noroviruses.[10] After successfully cloning the virus, the norovirus RNA genome was well characterized. Since then, more and more molecular biological methods for studying norovirus have been developed. Recently, the detection of noroviruses is majorly carried out using molecular nucleic acid-based methods because they are significantly superior in terms of speed, specificity, and sensitivity than other traditional methods. An array of molecular methods has been developed and applied for norovirus detection. The use of molecular diagnostic methods for norovirus detection is covered in more details in a recent review.[4]

8.6.4.1 REVERSE TRANSCRIPTION-PCR (RT-PCR)

In 1983, Dr. Kary Mullis at Cetus Corporation conceived of PCR. PCR is the enzymatic amplification of specific DNA sequences. This process uses multiple cycles of template denaturation, primer annealing, and primer elongation to amplify DNA sequences. PCR uses DNA as starting material. RT-PCR is used when the starting material is RNA. In this method,

RNA is first transcribed into complementary DNA (cDNA) by reverse transcriptase from total RNA or messenger RNA (mRNA). The cDNA is then used as the template for the PCR reaction. RT-PCR assays have been useful in detecting norovirus in foodborne outbreaks of gastroenteritis.[40]

8.6.4.2 REAL-TIME QUANTITATIVE PCR (RT-qPCR)

RT-qPCR assays are the preferred laboratory molecular method for detecting norovirus.[3] Traditional detection of amplified viral RNA relies upon gel electrophoresis. RT-qPCR is an advanced form of the RT-PCR. It is a major development of PCR technology that enables reliable detection and measurement of amplicons generated during each cycle of PCR process. A dual-labeled oligonucleotide probe which is designed to hybridize within the target sequence is generally introduced in RT-qPCR. Cleavage of the probe during PCR because of the 5' nuclease activity of *Taq* polymerase can be used to detect amplification of the target-specific products. These assays are very sensitive and can detect as few as 10–100 viral genome copies per reaction in food.[3,41] The method is able to differentiate genogroup I and II noroviruses using genogroup specific-primers. Table 8.1 lists some selected oligonucleotide primers and probes commonly used

TABLE 8.1 Oligonucleotide Primers and Probes Used for Detection and Genotype Identification of Norovirus (Source: Surveillance Network for Norovirus, USA).

Primer or probe name	RT-PCR target	Sequence, 5'—3'
TVN-L1	ORF2–ORF3	GGG TGT GTT GTG GTG TTG T26VN
L1	ORF2–ORF3	GGG TGT GTT GTG GTG TTG
EVP2F	P2 (GII.4 specific)	GTR CCR CCH ACA GTT GAR TCA
EVP2R	P2 (GII.4 specific)	CCG GGC ATA GTR GAY CTR AAG AA
Cap D1	Region D GII	TGT CTR STC CCC CAG GAA TG
Cap C	Region D GII	CCT TYC CAK WTC CCA YGG
Cap D3	Region D GII	TGY CTY ITI CCH CAR GAA TGG
Cog 2F	ORF1–ORF2 junction (GII)	CAR GAR BCN ATG TTY AGR TGG ATG AG
Cog 2R	ORF1–ORF2 junction (GII)	TCG ACG CCA TCT TCA TTC ACA
Ring 2	ORF1–ORF2 junction (GII)	Cy5-TGG GAG GGC GAT CGC AAT CT-BHQ
Ring 1C	ORF1–ORF2 junction (GI)	FAM-AGA TYG CGI TCI CCT GTC CA-BHQ
Cog 1F	ORF1–ORF2 junction (GI)	CGY TGG ATG CGI TTY CAT GA
Cog 1R	ORF1–ORF2 junction (GI)	CTT AGA CGC CAT CAT CAT TYA C

for detection and identification of norovirus in RT-qPCR assays.[3] With improved sensitivity, specificity, and a rapid turnaround time, RT-qPCR is most commonly used as a reference laboratory approach to detect norovirus in clinical specimens, food, and environmental samples.[20] In food samples, where the level of virus contamination is often very low, nested RT-qPCR was developed and offered a significant advantage in sensitivity to detect hepatitis A virus in green onions.[41] This nested RT-qPCR method can be potentially adopted as a regulatory method for norovirus detection in food.

8.6.4.3 BIO-PLEX SOLUTION HYBRIDIZATION

To enhance analytical capabilities on detection and identification of noroviruses, we have been developing a Bio-Plex method that could detect and genotype multiple pathogenic targets in a single assay. The technological platform also allows automation and can analyze up to 100 different assays multiplexed within each well of a micro plate. This Bio-Plex method can determine the noroviruses at genogroup levels. Data obtained demonstrated that the Bio-Plex method reached the same limit of detection as that of RT-qPCR.[42]

8.6.4.4 DNA MICROARRAYS

DNA microarray technology was first introduced as a new tool of molecular biology in 1995 for large-scale gene expression study at Stanford University by Schena and co-workers.[43] With this technology, it is possible to have hundreds and thousands of nucleic acid hybridization reaction to be performed simultaneously on a solid substrate. Microarray technology has been successfully used in many different experimental settings, such as gene expression profiling, genomic studies, and mutation analysis. When applied to foodborne viruses, specific sequences of viral pathogens are detected in the nucleic acids prepared from samples through hybridization to the known viral sequence probes printed on the microarray chips. Noroviruses display remarkably high genetic diversity which poses the challenge for accurate identification of the virus variants using RT-qPCR. A recent report showed that a single nucleotide polymorphism in the TaqMan probe-binding site resulted in decreased sensitivity of RT-qPCR for detection of a GII.4 norovirus variant strain, producing a false-negative result.[44] Broad range microarrays that cover known norovirus genotypes can offer high discriminatory power to identify genomic diversity of the virus strains through the use of linear sequence-independent amplification.[45,46] or random PCR[47] to enhance the limit of detection. Known viruses are detected on the microarray, as well as novel viruses with at least certain level of relatedness to the known ones.

8.6.4.5 DNA SEQUENCING

8.6.4.5.1 Traditional DNA Sequencing

DNA sequencing technology can be used to determine all or part of the nucleotide sequence of a norovirus genome. The DNA sequencing results in generating true base-by-base nucleotide information that provides the most accurate data for norovirus detection. It is also the most fundamental level of knowledge of norovirus genome. In order to characterize the norovirus strains in clinical, food, and environmental samples, the use of traditional sequencing methods, for example, Sanger sequencing to sequence the norovirus specific RT-PCR products is generally applied. For reliable determination of the genotype of a virus strain, sequencing the complete major capsid region of the norovirus genome is recommended.[17] Phylogenetic comparison of the nucleotide sequences obtained from food and patient samples may allow determining if the presence of norovirus in food samples is related to those of the patient samples, which can help trace the origin of the food contamination. DNA sequencing coupled with phylogenetic analysis provides a valuable tool for norovirus outbreak investigation and epidemiological studies.

8.6.4.5.2 Whole-Genome Sequencing

While traditional sequencing method has been used to routinely sequence the part or complete genomes of noroviruses, it is too time consuming. The recent development of whole-genome sequencing (WGS) technology (also referred to as "next-generation sequencing") provides an unprecedented way for rapid and high throughput sequencing of the genomes of known and unknown microbial organisms. For investigation of a foodborne outbreak, rapid identification and characterization of the pathogen should be a high priority. Once the relationship between a potential new causative infectious agent and previously known viruses is identified by WGS, epidemiological information can be used to plan and evaluate strategies to prevent the disease. In addition to norovirus detection and typing in epidemiological outbreak investigations,[48,49] this technology has been used for evolution studies to characterize norovirus communities and dynamics.[50,51] In general, WGS allows for highly improved discriminatory ability and determination of evolutionary association among isolates. The rapidly emerging use of WGS for norovirus identification will increase the power to detect norovirus-related foodborne disease outbreaks and help define pathogen sources throughout the food chain.

8.7 TREATMENT

Antiviral therapy is currently not available for treating norovirus illness. Because it is a viral (not bacterial) infection, norovirus illness cannot be treated with antibiotics.[3] The treatment of norovirus illness is supportive involving mainly the reversal of fluid and electrolyte balance. If patients have norovirus illness, they should drink adequate liquids to replace fluids lost from diarrhea or vomitus. This will help prevent dehydration which can lead to serious problems. Severe dehydration may require hospitalization for treatment with fluids given intravenously.[3]

8.8 PREVENTION

8.8.1 VACCINE

Vaccines are considered the most effective method of preventing infectious diseases.[52] Despite the pressing need to control and prevent norovirus outbreaks and illness, no licensed vaccines are currently available. The lack of appropriate model systems to explore norovirus pathogenesis and vaccine target efficacy has hampered the research for norovirus vaccine development.[53,54] An alternate approach has been focused on the production of norovirus VLPs in several expression systems using recombinant technology. VLPs can be produced by the expression of the major capsid protein VP1 in recombinant virus expression vector systems (e.g., baculovirus, Venezuelan equine encephalitis virus in insect cells), transgenic plants (tobacco, tomato, and potato), and yeast.[10,55–57] As stated above, recombinant VLPs are structurally and antigenically similar to their corresponding wild-type viruses and are highly immunogenic. The development of VLPs has played an important role in leading significant progress toward vaccine candidates designed to protect against norovirus infection. The VLPs-based vaccination has shown the protection against homologous viral challenge in a human study.[58] A number of norovirus vaccine candidates including bivalent and trivalent VLPs have been evaluated in preclinical trials.[54]

Recent years have seen that the establishment of cell culture systems for the replication of human noroviruses is on the horizon.[59,60] These tools will facilitate the studies of pathogenesis and immunology of norovirus and hopefully advance the development of improved vaccines.

8.8.2 PERSONAL HYGIENE

Since currently no antiviral therapy or vaccine is available for treatment or prevention of norovirus illness, prevention of norovirus infection relies mainly on strict community, and personal hygiene measures. Viruses are transmitted directly from person to person and indirectly via contaminated water and food. An effective way to prevent norovirus infection is to practice proper hand washing and general cleanliness.[3] Hand washing can reduce the spread of pathogenic microorganisms that are transmitted through food. Hands should be washed carefully with soap and water especially after using the toilet and changing diapers, and always before eating, preparing, or handling food. Noroviruses can be found in patient's vomit or stool even before they start feeling sick. The virus can survive in infected individual's stool for more than 2 weeks. Thus, it is important to continue washing hands often during this period of time. Wash fruits and vegetables and cook seafood thoroughly. Carefully wash fruits and vegetables before preparing and eating them. Be aware that noroviruses are relatively resistant. They can survive temperatures as high as 140°F.[3] Keep sick infants and children out of areas where food is being handled and prepared. This also applies to sick workers in settings, such as schools, daycares, and cruise ships where they may expose people to norovirus. More information about the personal hygiene can be found from FDA

Employee Health and Personal Hygiene Handbook (http://www.fda.gov/Food/FoodSafety/RetailFoodProtection/IndustryandRegulatoryAssistanceandTrainingResources/ucm113827.htm).

8.8.3 DISINFECTION

Good sanitation can reduce the incidence of norovirus contamination. Cleaning and disinfection of contaminated surfaces are also important. After throwing up or having diarrhea, immediate cleaning and disinfection of contaminated surfaces are needed.[3] There are diverse ways to disinfect the viruses in food, such as cooking, heating irradiation, desiccation, depuration, and chemical disinfection. Thorough cooking is one of the best and most practical methods to totally inactivate norovirus in food.[61]

8.9 LESSONS LEARNED FROM NOROVIRUS OUTBREAKS AND SURVEILLANCE INFORMATION

8.9.1 FOOD SERVICE

A broad range of foods (e.g., shellfish) fresh produce and ready-to-eat/catered foods has been implicated in norovirus foodborne outbreaks. The outbreaks from contaminated food are common in food service settings, such as restaurants, catering or banquet facility, and cruise ships. These are the places where people often eat food processed and prepared by food service workers. Foods can get contaminated with norovirus when infected food workers who have stool or vomit on their hands touch ready-to-eat foods or handle after being cooked foods before serving them. Infected food workers cause about 70% of reported norovirus outbreaks from contaminated food.[3] As food service has a key role in preventing norovirus outbreaks, some problems are evident in it; for example, food service workers infected with norovirus are highly contagious, and they often go to work when they are sick. In addition, norovirus is resilient to environment stress and remain infectious on food even at freezing temperatures and until heated above 140°F. Food service industry should keep vigilant on reducing norovirus food contamination. Food safety training should be enhanced to ensure that all food service workers follow food safety practices outlined in the FDA model food code and CDC guidelines. Some ways that food service workers can follow to prevent the virus spreading include: (1) do not prepare food while sick and for at least 48 h after symptoms stop; (2) avoid touching food with bare hands; (3) wash hands often with soap and water; (4) carefully rinse fruits and vegetables, and cook shellfish fully; (5) clean and sanitize kitchen counters, surfaces, and utensils regularly; and (6) wash table linens and other laundry thoroughly.[3]

8.9.2 CRUISE SHIPS

Over the past years, a global increase in the number of norovirus outbreaks on cruise ships was noticed (Table. 8.2). Noroviruses are transmitted by the fecal-oral route either directly from person-to-person contact or indirectly through consumption of contaminated foods. According to Cruise Lines International Association (CLIA), one of the cruise industries' top priorities is to prevent gastrointestinal illness from being brought on board a ship. In fact, norovirus outbreaks are found and reported more frequently on a cruise ship than on land. The reasons for why noroviruses are often associated with cruise ships are that ships may increase the amount of group contact and people joining the ship may bring the virus to other passengers and crews. Virus might also be introduced when ships dock in countries where sanitation is poor, either through contaminated food or passengers becoming infected while embarking the ships. Also, outbreaks of norovirus usually occur where large volumes of foods are prepared at one time. For epidemic control, the Vessel Sanitation Program (VSP) at the CDC assists the cruise ship industry to prevent and control the introduction, transmission, and spread of gastrointestinal illnesses on cruise ships.[3] Cruise ships participated this program are required to report the total number of gastrointestinal illness cases before the ship arrives at a United States port when sailing from a foreign port. Under the authority of the Public Health Service Act (42 U.S.C. Section 264 Quarantine and Inspection Regulations to Control Communicable Diseases), VSP will inspect cruise ships in periodic, unannounced operational sanitation inspections, monitor gastrointestinal illnesses, investigate outbreaks, train the cruise ship employees on public health practices, provide health education to the cruise ship industry, the traveling public, public health professionals, state and local health authorities, and the media.[3]

TABLE 8.2 Outbreak Updates for International Cruise Ships (Source: Centers for Disease Control and Prevention 2014).

Cruise line	Cruise ship	Sailing dates	Causative agent
	2014		
Princess Cruises	Crown Princess	10/18–11/16	Norovirus
Princess Cruises	Crown Princess	4/5–4/12	Norovirus and Enterotoxigenic *E. coli* (ETEC)
Royal Caribbean Cruise Line	Grandeur of the Seas	4/5–4/12	Norovirus
Royal Caribbean Cruise Line	Grandeur of the Seas	3/28–4/5	Norovirus
Holland America Line	MS Maasdam	3/2–3/28	Unknown
Holland America Line	MS Veendam	2/8–2/22	Norovirus
Princess Cruises	Caribbean Princess	1/25–2/1	Norovirus
Royal Caribbean Cruise Line	Explorer of the Seas	1/21–1/31	Norovirus
Norwegian Cruise Line	Norwegian Star	1/5–1/19	Norovirus
	2013		
Norwegian Cruise Line	Norwegian Gem	11/16–11/25	Norovirus
Celebrity Cruise Lines	Celebrity Summit	9/21–10/5	Norovirus
Celebrity Cruise Lines	Celebrity Millennium	04/25–05/10	Norovirus
Crystal Cruises	Crystal Symphony	04/29–05/06	Norovirus
Holland America Line	Veendam	04/13–05/04	*Enterotoxigenic E. coli* (ETEC)
Celebrity Cruises	Celebrity Solstice	04/08–04/25	Norovirus
Celebrity Cruises	Celebrity Infinity	03/17–04/01	Norovirus
Princess Cruises	Ruby Princess	03/03–03/10	Norovirus
Royal Caribbean Cruise Line	Vision of the Seas	02/25–03/08	Norovirus
	2012		
Cunard Line	Queen Mary 2	12/22–01/03	Norovirus
Princess Cruises	Emerald Princess	12/17–12/27	Norovirus
Prestige Cruise Holdings	Oceania Riviera	11/15–11/29	Norovirus
Holland America Line	Amsterdam	11/11–12/5	Norovirus
Princess Cruises	Ruby Princess	10/09–10/28	Norovirus and *Enterotoxigenic E. coli* (ETEC)
Princess Cruises	Dawn Princess	08/21–09/13	Norovirus
Royal Caribbean Cruise Line	Rhapsody of the Seas	08/24–08/31	Norovirus
Carnival Cruise Line	Carnival Glory	08/06–08/11	Norovirus
Princess Cruises	Sun Princess	07/08–07/21	Norovirus
Princess Cruises	Ruby Princess	02/26–03/04	Norovirus

TABLE 8.2 *(Continued)*

Cruise line	Cruise ship	Sailing dates	Causative agent
Princess Cruises	Crown Princess	02/04–02/09	Norovirus
Celebrity Cruises	Celebrity Silhouette	01/29–02/10	Norovirus
Celebrity Cruises	Celebrity Constellation	01/28–02/11	Norovirus
Princess Cruises	Crown Princess	01/28–02/04	Norovirus
P & O Cruises	Aurora	01/04–01/26	Norovirus
Royal Caribbean Cruise Line	Voyager of the Seas	01/28–02/04	Norovirus
	2011		
Holland America Line	MS Maasdam	12/23–01/02	Norovirus
Norwegian Cruise Line	Norwegian Spirit	12/18–12/25	Norovirus
Holland America Line	Ryndam	12/11–12/18	Norovirus
Celebrity Cruises	Celebrity Solstice	11/28–12/11	Norovirus
Holland America Line	Ryndam	11/03–11/20	Norovirus
Princess Cruises	Sea Princess	05/30–06/09	Norovirus
Lindblad Expeditions	National Geographic Sea Lion	05/21–05/28	Unknown
Princess Cruises	Sea Princess	05/20–05/30	Norovirus
Celebrity Cruises	Celebrity Millennium	05/20–05/27	Unknown
Princess Cruises	Coral Princess	05/04–05/19	Norovirus
Princess Cruises	Coral Princess	04/24–05/04	Enterotoxigenic *E. coli* (ETEC)
Oceania Cruises	MV Marina	02/26–03/15	Enterotoxigenic *E. coli* (ETEC)
Celebrity Cruises	Celebrity Solstice	01/30–02/06	Norovirus
Royal Caribbean Cruise Line	Radiance of the Seas	01/03–01/08	Norovirus
	2010		
Cunard Line	Queen Mary 2	12/19–01/03/2011	Unknown
Crystal Cruises	Crystal Symphony	11/02–11/21	Unknown
Holland America Line	Nieuw Amsterdam	10/18–11/07	Norovirus
Carnival Cruise Lines	Carnival Glory	10/09–10/16	Norovirus
Holland America Line	Zuiderdam	04/20–05/08	Unknown
Celebrity Cruises	Mercury	03/08–03/19	Norovirus
Celebrity Cruises	Mercury	02/26–03/08	Norovirus
Royal Caribbean International	Jewel of the Seas	02/22–03/05	Unknown
Celebrity Cruises	Millennium	02/22–03/05	Norovirus
Holland America Line	Maasdam	02/19–03/05	Norovirus

TABLE 8.2 *(Continued)*

Cruise line	Cruise ship	Sailing dates	Causative agent
Celebrity Cruises	Mercury	02/15–02/26	Norovirus
Fred Olsen Cruise Lines	Balmoral	01/05–02/04	Unknown
Cunard Cruise Line	Queen Victoria	01/12–01/27	Unknown
Cunard Cruise Line	Queen Victoria	01/04–01/12	Norovirus
	2009		
Holland America Line	Amsterdam	12/02–12/23	Norovirus
Regent Seven Seas Cruises	Seven Seas Mariner	12/02–12/20	Norovirus
Norwegian Cruise Line	Norwegian Spirit	10/23–11/08	Norovirus
Holland America Line	Noordam	10/16–11/1	Norovirus
Holland America Line	Amsterdam	5/23–5/30	Unknown
Princess Cruises	Coral Princess	5/18–5/25	Unknown
Princess Cruises	Island Princess	4/23–5/07	*Shigella sonnei* and Enterotoxigenic *E. coli* (ETEC)
Carnival Cruise Line	Carnival Liberty	4/18–4/25	Norovirus
Holland America Line	Amsterdam	4/2–4/24	*Cyclospora cayetanensis*
Celebrity Cruises	Celebrity Mercury	2/28–3/15	Norovirus
Holland America Line	Oosterdam	2/21–2/28	Norovirus
Holland America Line	Zaandam	2/5–2/20	Norovirus
Princess Cruises	Coral Princess	2/3–2/13	Norovirus
Celebrity Cruises	Celebrity Mercury	1/3–1/17	Norovirus
Holland America Line	Maasdam	1/2–1/9	Unknown
	2008		
Holland America Line	Zuiderdam	10/23–11/09	Norovirus
Princess Cruises	Caribbean Princess	10/19–10/26	Norovirus
Norwegian Cruise Line	Norwegian Jewel	10/8–10/18	Norovirus
American Canadian Cruise Line	Grande Caribe	9/30–10/12	Norovirus
Norwegian Cruise Line	Norwegian Jewel	9/28–10/8	Norovirus
Regent Seven Seas Cruises	Seven Seas Mariner	8/27–9/3	Norovirus
Holland America Line	Veendam	6/13–6/20	Norovirus
Norwegian Cruise Line	Norwegian Dream	4/13–5/1	Norovirus and Enterotoxigenic *E. coli* (ETEC)
Holland America Line	Zaandam	3/19–4/3	Norovirus

TABLE 8.2 *(Continued)*

Cruise line	Cruise ship	Sailing dates	Causative agent
Holland America Line	Ryndam	2/15–2/25	Norovirus
Carnival Cruise Line	Holiday	1/24–1/28	Norovirus
Norwegian Cruise Lines	Norwegian Star	1/19–1/27	Norovirus
Holland America Line	Noordam	1/5–1/16	Norovirus
Holland America Line	Volendam	1/2–1/12	Norovirus
Princess Cruises	Pacific Princess	12/15–1/10	Enterotoxigenic *E. coli* (ETEC)
	2007		
Princess Cruises	Emerald Princess	12/16–12/26	Norovirus
Norwegian Cruise Line	Norwegian Pearl	11/30–12/9	Norovirus
Holland America Line	Zaandam	11/5–11/20	Norovirus
Norwegian Cruise Line	Norwegian Dawn	10/14–10/26	Norovirus
Norwegian Cruise Line	Norwegian Crown	8/12–8/19	Norovirus
Carnival Cruise Line	Carnival Spirit	6/13–6/20	Norovirus
Royal Caribbean Cruise Line	Liberty of the Seas	5/26–6/2	Norovirus
Norwegian Cruise Line	Norwegian Star	5/19–5/26	Norovirus
Norwegian Cruise Line	Norwegian Pearl	4/13–4/22	Specimens not obtained
Holland America Line	Volendam	4/12–4/22	Specimens not obtained
Regent Seven Seas Cruises	Seven Seas Navigator	4/5–4/16	Norovirus
Celebrity Cruises	Zenith	4/2–4/13	Norovirus
Majestic America Line	Empress of the North	3/16–3/21	Norovirus
Princess Cruises	Regal Princess	3/13–3/25	Unknown
Regent Seven Seas Cruises	Seven Seas Mariner	2/25–3/9	*Shigella, Salmonella, Enterobacter,* and *Entameoba histolytica*
Holland America	Ryndam	3/3–3/13	Norovirus
		2/21–3/3	Norovirus
		1/22–2/21	Norovirus
Holland America Line	Westerdam	2/2–2/11	Norovirus
Princess Cruises	Island Princess	1/26–2/10	Norovirus
Cruise West	Spirit of Nantucket	1/27–2/3	Unknown
Holland America Line	Volendam	1/22–2/1	Specimens not obtained
Cunard Line	Queen Elizabeth 2	1/8–1/22	Norovirus

TABLE 8.2 *(Continued)*

Cruise line	Cruise ship	Sailing dates	Causative agent
	2006		
Princess Cruises	Regal Princess	12/7–12/27	Norovirus
Princess Cruises	Sun Princess	11/30–12/10	Norovirus
Norwegian Cruise Line	Norwegian Wind	11/29–12/9	Norovirus
Holland America Line	Zaandam	11/26–12/11	Norovirus
Royal Caribbean Cruise Line	Freedom of the Seas	11/26–12/3/06	Norovirus
		12/3–12/10	Norovirus
Royal Caribbean Cruise Line	Serenade of the Seas	11/18–11/25	Norovirus
Celebrity Cruises	Summit	11/11–11/25	Unknown, negative for norovirus
Royal Caribbean Cruise Line	Adventure of the Seas	11/5–11/12	Norovirus
Carnival Cruise Line	Carnival Liberty	11/3–11/19	Norovirus
Royal Caribbean Cruise Line	Serenade of the Seas	10/29–11/11	Norovirus
Princess Cruises	Island Princess	10/21–11/5	Norovirus
Celebrity Cruises	Constellation	10/27–11/7	Norovirus
Celebrity Cruises	Constellation	10/16–10/27/06	Norovirus
Princess Cruises	Diamond Princess	9/2–9/9	Norovirus
Royal Caribbean Cruise Line	Radiance of the Seas	9/2–9/9	Unknown, negative for norovirus
Royal Caribbean Cruise Line	Mariner of the Seas	7/23–7/30	Norovirus
Celebrity Cruises	Infinity	7/9–7/16	Norovirus
Celebrity Cruises	Mercury	6/2–6/9	Norovirus
Holland America Line	Ryndam	5/28–6/4	Norovirus
Holland America Line	Ryndam	5/14–5/21	Norovirus
Norwegian Cruise Line	Norwegian Star	5/14–5/21	Norovirus
Norwegian Cruise Line	Norwegian Sun	5/13–5/20	Norovirus
Holland America Line	Veendam	4/22–5/9	Norovirus
Royal Caribbean Cruise Line	Explorer of the Seas	4/23–4/30	Norovirus
Princess Cruises	Island Princess	3/7–3/22	Norovirus
Holland America Line	Amsterdam	3/7–3/22	Norovirus
		3/22–4/6	Norovirus
Celebrity Cruises	Mercury	3/6–3/17	Norovirus
		3/17–3/27	Norovirus
Royal Caribbean Cruise Line	Grandeur of the Seas	3/6–3/11	Norovirus
Holland America Line	Volendam	3/3–3/13	Norovirus
Regent Seven Seas Cruises	Seven Seas Mariner	2/28–3/13	Unknown
Royal Caribbean Cruise Line	Explorer of the Seas	2/26–3/5	Norovirus

TABLE 8.2 *(Continued)*

Cruise line	Cruise ship	Sailing dates	Causative agent
Cunard Line	Queen Mary 2	2/8–2/22	Enterotoxigenic *E. coli* (ETEC)
P & O Cruises	Minerva II	1/12–1/26	Unknown
Holland America Line	Zuiderdam	12/30/05–1/7/06	Norovirus
	2005		
Royal Caribbean Cruise Line	Splendour of the Seas	11/12–11/26	Norovirus
Princess Cruises	Sun Princess	5/9–5/16	Norovirus
Princess Cruises	Dawn Princess	4/29–5/14	Norovirus
Celebrity Cruises	Horizon	4/23–4/30	Unknown
Crystal Cruises	Crystal Symphony	4/14–4/25	Unknown
Norwegian Cruise Line	Norwegian Crown	3/13–3/28	Unknown
Princess Cruises	Royal Princess	3/7–3/24	Unknown
Holland America Line	Veendam	3/5–3/12	Norovirus
Celebrity Cruises	Zenith	2/27–3/13	Norovirus
Carnival Cruise Line	Celebration	2/21–2/26	Norovirus
		2/17–2/21	Norovirus
Regent Seven Seas Cruises	Seven Seas Mariner	2/12–2/24	*Salmonella*
Royal Caribbean Cruise Line	Empress of the Seas	1/17–1/28	Norovirus
Royal Caribbean Cruise Line	Mariner of the Seas	1/16–1/23	Norovirus
Holland America Line	Ryndam	1/13–1/29	Norovirus
Princess Cruises	Sun Princess	1/8–1/18	Norovirus
Holland America Line	Veendam	1/3–1/15	Norovirus
		1/15–1/29	Norovirus
Royal Caribbean Cruise Line	Enchantment of the Seas	1/3–1/8	Norovirus
	2004		
Princess Cruises	Coral Princess	12/28–1/07	Norovirus
Princess Cruises	Golden Princess	12/12–12/19	Norovirus
SilverSea Cruise Line	Silver Shadow	12/4–12/14	Specimens not obtained
Norwegian Cruise Line	Norwegian Dream	12/5–12/12	Norovirus
Institute for Shipboard Education	M/V Explorer	6/19–12/8	Unknown
Holland America Line	Veendam	9/26–10/14	Unknown
Princess Cruises	Regal Princess	9/11–9/20	Unknown

TABLE 8.2 *(Continued)*

Cruise line	Cruise ship	Sailing dates	Causative agent
Princess Cruises	Sun Princess	9/13–9/20	Unknown
		9/6–9/13	Norovirus
Clipper Cruise Line	Nantucket Clipper	9/3–8/11	Specimens not obtained
SilverSeas Cruise Line	Silver Shadow	7/26–8/6	Unknown
Princess Cruises	Pacific Princess	7/25–8/17	Unknown
Royal Caribbean Cruise Line	Serenade of the Seas	6/6–6/13	Unknown
Princess Cruises	Island Princess	6/5–6/12	Unknown
		5/29–6/5	Norovirus
Norwegian Cruise Line	Norwegian Crown	5/2–5/9	Norovirus
		4/17–5/2	Norovirus
Holland America Line	Veendam	4/17–5/2	Norovirus
Norwegian Cruise Line	Norwegian Wind	4/17–5/2	Specimens not obtained
Holland America Line	Rotterdam	4/2–4/15	Unknown
Holland America Line	Amsterdam	3/31–4/17	Enterotoxigenic *E. coli* (ETEC)
Celebrity Cruises	Zenith	3/28–4/11	Norovirus
Celebrity Cruises	Horizon	3/22–4/2	Norovirus
Norwegian Cruise Line	Norwegian Crown	3/22–4/4	Norovirus
		3/7–3/22	Unknown, negative for norovirus
Princess Cruises	Royal Princess	3/8–3/25	Unknown, negative for norovirus
Cunard Line	Queen Mary 2	2/23–3/6	Norovirus
Royal Caribbean Cruise Line	Brilliance of the Seas	2/16–2/27	Norovirus
Cunard Line	Queen Elizabeth 2	2/17–3/1	Unknown
ResidenSea Cruise Line	The World	2/3–2/24	Specimens not obtained
P & O Cruises	Oriana	2/20–3/5	Norovirus
Carnival Cruise Line	Carnival Celebration	2/9–2/14	Norovirus
Holland America Line	Ryndam	2/7–2/14	Norovirus
Princess Cruises	Coral Princess	1/12–1/22	Norovirus
Cunard Line	Queen Elizabeth 2	1/5–1/19	Specimens not obtained
Windjammer Barefoot Cruises	Legacy	1/4–1/10	Specimens not obtained

8.9.3 REGULATORY REQUIREMENTS

A wholesome and safe food supply plays an important role in public health. Through science-based policies and practices, the regulatory agencies craft the best ways forward in meeting the foodborne challenges. In 2014, CDC published the most recent norovirus recommendations which provide guidelines for outbreak management and disease prevention. Specific recommendations include standardized collection of clinical specimens during norovirus outbreaks, use of several recently launched surveillance systems for reporting of norovirus outbreaks, and use of appropriate control measures focusing on hand hygiene and environmental disinfection. Their implementation by public health professionals is intended to guide efficient use of public health resources for effective prevention and control of norovirus disease. Several national and international regulatory surveillance systems/programs for norovirus are listed as follows.

8.9.3.1 VESSEL SANITATION PROGRAM

In recent years, noroviruses are the primary cause of epidemic viral gastroenteritis and the leading cause of foodborne outbreaks in cruise ships. Developing strategies to reduce norovirus survival have been an ongoing challenge. According to CDC, cruise ship outbreak updates are posted when they meet the following criteria: (1) Fall within the purview of VSP; (2) Sailing on voyages from 3 to 21 days; (3) Carrying 100 or more passengers; (4) Cruise ships in which 3% or more of passengers or crew reported symptoms of diarrheal disease to the ships medical staff during the voyage; and (5) Gastrointestinal illness outbreaks of public health significance.

Passengers and crews aboard cruise ships are affected frequently by norovirus outbreaks.[2] Virus is generally introduced on board by passengers or crew infected before embarkation but might also result from food items contaminated before loading or persistently contaminated environmental surfaces from previous cruises. CDC's VSP will assist the cruise ship industry in preventing and controlling the introduction and transmission of gastrointestinal illness by inspecting cruise ships, monitoring gastrointestinal illnesses, and responding to outbreaks (http://www.cdc.gov/nceh/vsp).

8.9.3.2 NATIONAL OUTBREAK REPORTING SYSTEM (NORS)

For epidemic control, CDC developed the NORS in 2009. The web-based platform supports outbreak reporting by partners in state, local, and territorial public health agencies. It provides the information about all foodborne and waterborne disease outbreaks and enteric disease transmitted by contact with environmental sources, infected individuals or animals, or unknown modes of transmission to CDC (Fig. 8.3). The continual improvement of the NORS is critical in helping to integrate and streamline surveillance system, enhance state and local health department outbreak reporting, and provide information to government agencies for prevention of future outbreaks.

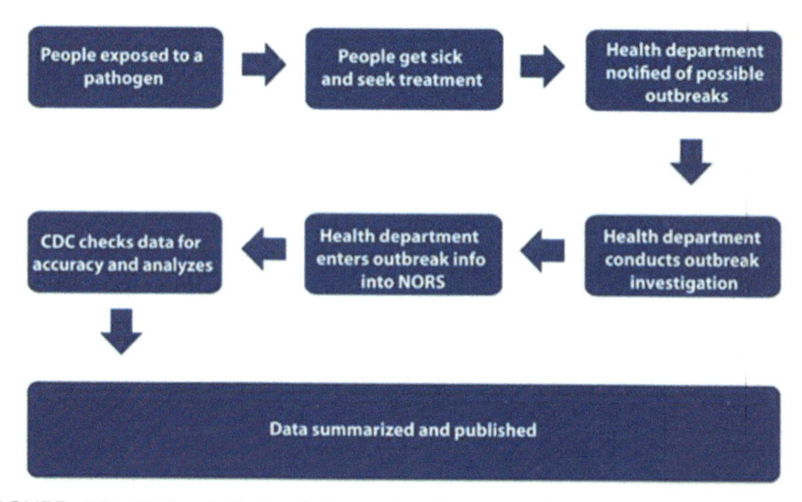

FIGURE 8.3 National Outbreak Reporting System (NORS) The flow of outbreak investigation: (1) people are exposed to a pathogen; (2) people get sick and seek treatment; (3) health department is notified of a possible outbreak; (4) health department conducts an outbreak investigation; (5) health department enters outbreak information into NORS; (6) CDC checks data for accuracy and analyses; and (7) data are summarized and published. (Source: Centers for Disease Control and Prevention)

8.9.3.3 NATIONAL NOROVIRUS OUTBREAK SURVEILLANCE NETWORK (CALICINET)

CaliciNet (Fig. 8.4) is a national norovirus outbreak surveillance network of federal, state, and local public health laboratories in the United States.

To increase the quality of national norovirus surveillance, CDC launched CaliciNet in 2009 to collect information on norovirus strains associated with gastroenteritis outbreaks in the United States. Public health laboratories electronically submit laboratory data, including genetic sequences of norovirus strains, and epidemiology data from norovirus outbreaks to the CaliciNet database. This network allows public health agencies to define which samples are parts of the same outbreak. The norovirus strains can be compared with other norovirus strains in the database, helping CDC link outbreaks to a common food source, monitor norovirus strains that are circulating, and identify newly emerging norovirus strains. CaliciNet aims to improve the network and provide a better understanding of norovirus and to help control its spread.[3]

FIGURE 8.4 National norovirus outbreak surveillance network. (Source: CaliciNet)

8.9.3.4 NOROVIRUS SENTINEL TESTING AND TRACKING NETWORK (NOROSTAT)

NoroSTAT was established as a collaborative network of nine state health departments and CDC working together to maintain standard practices for norovirus outbreak surveillance in 2012. It aims to improve the timeliness, completeness, and consistency of norovirus outbreak reporting. The system allows CDC to rapidly link epidemiologic and laboratory data submitted to CDC's CaliciNet and NORS by participating state health departments.

8.9.3.5 OTHER SURVEILLANCE SYSTEMS

Foodborne diseases active surveillance network (FoodNet) and new vaccine surveillance network (NVSN) are used by CDC to do norovirus testing on patients with gastroenteritis. These systems help generate new estimates of norovirus illness and monitor trends over time.

NoroNet is an informal network that virologists and epidemiologists working in public health institutes or university in 13 European countries share surveillance and research data on enteric viruses, focusing mainly on norovirus. The network aims primarily to enlarge the knowledge on geographical and temporal trends in the emergence and spread of norovirus variants, thus limiting the impact and scale of future norovirus epidemics.

ACKNOWLEDGMENTS

The use of the commercial names or products here is solely for the purpose of providing scientific information. No official support or endorsement of this article by the FDA is intended or should be inferred. The authors would like to thank Dr. Ivica Arsov (York College) for critically reading the chapter.

KEYWORDS

- **norovirus**
- **molecular detection**
- **RT-qPCR**
- **WGS**
- **prevention**

REFERENCES

1. Zahorsky, J. *Arch. Pediatr.* **1929,** *46,* 391–395.
2. Kapikian, A. Z.; Wyatt, R. G; Dolin, R.; Thornhill, T. S.; Kalica, A. R.; Chanock, R. M. *J. Virol.* **1972,** *10,* 1075–1081.

3. CDC. http://www.cdc.gov/norovirus/ 2016.
4. Chen, H.; Hu, Y. *Open Microbiol. J.* 2016, *10,* 78–89.
5. Chen, H. *Nucleic Acids: From Basic Aspects to Laboratory Tools*; InTechOpen: London, 2016; Vol. 3, p 37.
6. Wobus, C. E.; Karst, S. M.; Thackray, L. B.; Chang, K. O.; Sosnovtsev, S. V.; Belliot, G.; Krug, A.; Mackenzie, J. M.; Green, K. Y.; Virgin, H. W. *PLoS Biol.* **2004,** *2,* e432.
7. Thorne, L. G.; Goodfellow, I. G. *J. Gen. Virol.* **2014,** *95,* 278–291.
8. Wobus, C. E.; Thackray, L. B.; Virgin, H. W. *J. Virol.* **2006,** *80,* 5104–5112.
9. Orchard, R. C.; Wilen, C. B.; Doench, J. G.; Baldrige, M. T.; McCune, B. T.; Lee, Y. C.; Lee, S.; Pruett-Miller, S. M.; Nelson, C. A.; Fremont, D. H. *Science* **2016,** *353,* 933–936.
10. Jiang, X.; Wang, M.; Graham, D. Y.; Estes, M. K. *J. Virol.* **1992,** *66,* 6527–6532.
11. Green, K. Y.; Chanock, R. M.; Kapikian, A. Z. Human Caliciviruses. In *Fields Virology,* 4th ed.; Knipe, D. M., Howley, P. M., Eds.; Lippincott Williams & Wilkins: New York, 2001; Vol. 27, p 841.
12. Kaplon, J.; Guenau, E.; Asdrubal, P.; Pothier, P.; Ambert, B. K. *Emer. Infect. Dis.* **2011,** *17,* 1120.
13. Adler, J. L.; Zickl, R. *J. Infect. Dis.* **1969,** *119,* 668–673.
14. Thornhill, T. S.; Wyatt, R. G.; Kalica, A. R.; Dolin, R.; Chanock, R. M.; Kapikian, A. G. *J. Infect. Dis.* **1977,** *135,* 20–27.
15. Appleton, H.; Buckley, M.; Thom, B. T.; Cotton, J. L.; Henderson, S. *Lancet* **1977,** *1,* 409–411.
16. Qu, L.; Murakami, K.; Broughman, J. R.; Lay, M. K.; Guix, S.; Tenge, V. R.; Atmar, R. L.; Estes, M. K. *J. Virol.* **2016,** *90,* 8906–8923.
17. Zheng, D. P.; Ando, T.; Fankhauser, R. L.; Beard, R. S.; Glass, R. I.; Monroe, S. S. *Virology* **2006,** *346,* 312–323.
18. Martella, V.; Lorusso, E.; Decaro, N., et al. *Emerg. Infect. Dis.* **2008,** *14,* 1306–1308.
19. Tse, H.; Lau, S. K.; Chan, W. M.; Choi, G. K.; Woo, P. C.; Yuen, K. Y. *J. Virol.* **2012,** *86,* 9531–9532.
20. Glass, R. I.; Parashar, U. D.; Estes, M. K. *N. Engl. J. Med.* **2009,** *361,* 1776–1785.
21. Siebenga, J. J.; Vennema, H.; Zheng, D. P., et al. *J. Infect. Dis.* **2009**; *200,* 802–812.
22. Lu, J.; Sun, L.; Fang, L.; Yang, F.; Mo, Y.; Lao, J; Zheng, H.; Tan, X.; Lin, H.; Rutherford, S.; Guo, L.; Ke, C.; Hui, L. *Emerg. Infect. Dis.* **2015,** *21,* 1240–1242.
23. Parra, G.; Green, K. *Emerg. Infect. Dis.* **2015,** *21,* 1477–1479.
24. de Graaf, M.; van Beek, J.; Vennema, H.; Podkolzin, A. T.; Hewitt, J.; Bucardo, F., et al. *Euro Surveill.* **2015,** *20,* 21178.
25. Tan, M.; Jiang, X. *Trends Microbiol.* **2011,** *19,* 382–388.
26. Vicente, N. C.; Allen, D.; Rodríguez-Díaz, J.; Iturriza-Gómara, M.; Buesa, J. *Virol. J.* **2016,** *13,* 82.
27. Green, K. Y. *Caliciviridae*: The Noroviruses. In *Fields Virology;* Knipe, D. M., Howley, P. M., Eds.; Lippincott Williams & Wilkins: Philadelphia, PA, 2007; Vol. 1, pp 949–979.
28. Teunis, P. F.; Moe, C. L.; Liu, P.; Miller, S. E.; Lindesmith, L.; Baric, R. S.; Le Pendu, J.; Calderon, R. L. *J. Med. Virol.* **2008,** *80,* 1468–1476.
29. Widdowson, M. A.; Monroe, S. S.; Glass, R. I. *Emerg. Infect. Dis.* **2005,** *11,* 735–737.

30. Donaldson, E. F.; Lindesmith, L. C.; Lobue, A. D.; Baric, R. S. *Nat. Rev. Microbiol.* **2010,** *8,* 231–241.
31. Vega, E.; Barclay, L.; Gregoricus, N.; Williams, K.; Lee, D.; Vinje, J. *Emerg. Infect. Dis.* **2011,** *17,* 1389–1395.
32. Vega, E.; Barclay, L.; Gregoricus, N.; Shirley, S. H.; Lee, D.; Vinjé, J. *J. Clin. Microbiol.* **2014,** *52,*147–155.
33. Verhoef, L.; Hewitt, J.; Barclay, L.; Ahmed, S.; Lake, R.; Hall, A. J.; Lopman, B.; Kroneman, A.; Vennema, H.; Vinje, J.; Koopmans, M. *Emerg. Infect. Dis.* **2015,** *21,* 592–599.
34. Richards, G. P. *J. Indust. Microb. Biotech.* **2001,** *27,* 117.
35. Lou, F.; Huang, P.; Neetoo, H.; Gurtler, J. B.; Niemira, B. A.; Chen, H.; Jiang, X.; Lia, J. *Appl. Environ. Microbiol.* **2012,** *78,* 5320–5327.
36. Rutjen, S. A.; Verschoor, F. R.; Poel, M. V.; Duijnhoven, Y. D.; Husman, A. R. *J. Food Protec.* **2006,** *69,* 1949–1956.
37. Croci, L.; Dubois, E.; Cook, N.; Medici, D.; Schultz, A. C.; China, B.; Rutjes, S. A.; Hoorfar, J.; Poel, W. *Food Anal. Methods.* **2008,** *1,* 73.
38. Kapikian, A. Z.; Chanock, R. M. Norwalk Group of Viruses. In *Virology,* 2nd ed.; Fields, B. N., Knipe, D. N., Eds.; Raven Press: New York, NY, 1990; pp 673–693.
39. Okitsu-Negishi, S.; Okame, M.; Shimizu, Y.; Phan, T. G.; Tomaru, T.; Kamijo, S.; Sato, T.; Yagyu, F.; Muller, W. E.; Ushijima, H. *J. Clin. Microb.* **2006,** *44,* 3784–3786.
40. Daniels, N. A.; Bergmire-Sweat, D. A.; Schwab, K. J.; Hendricks, K. A.; Reddy, S.; Rowe, S. M.; Fankhauser, R. L.; Monroe, S. S.; Atmar, R. L.; Glass, R. I.; Mead, P. *J. Infect. Dis.* **2000,** *181,* 1467–1470.
41. Hu, Y.; Arsov, I. *Lett. Appl. Microb.* **2009,** *49,* 615–619.
42. Hu, Y.; Chou, K.; Chen, H.; Khanna, M.; Orlandi, P.; Williams-Hill, D. In *The Bioplex Working Group of the ORA Virology Initiative*; FDA Sciences Conference: Silver Spring, MD, 2012.
43. Schena, M.; Shalon, D.; Davis, R. W.; Brown, P. O. *Science.* **1995,** *270,* 467–470.
44. Zhuo, R.; Hasing, M. E.; Team of Molecular Diagnostics; Pang, X. *J. Clin Microbiol.* **2015,** *53,* 3353–3354.
45. Chen, H.; Chen, X.; Hu, Y.; Yan, H. *Appl. Microbiol. Biotech.* **2013,** *97,* 4129–4139.
46. Hu, Y.; Yan, H.; Mammel, M.; Chen, H. *AMB Express.* **2015,** *5,* 69.
47. Martínez, M. A.; Soto-Del Río Mde, L.; Gutiérrez, R. M.; Chiu, C. Y.; Greninger, A. L.; Contreras, J. F.; Lopez, S.; Arias, C. F.; Isa, P. *J. Clin. Microbiol.* **2015,** *53,*136–145.
48. Kundu, S.; Lockwood, J.; Depledge, D. P.; Chaudhry, Y.; Aston, A.; Rao, K.; Hartley, J. C.; Goodfellow, I.; Breuer, J. *Clin. Infect. Dis.* **2013,** *57,* 407–414.
49. Wong, T. H.; Dearlove, B. L.; Hedge, J.; Giess, A. P.; Piazza, P.; Trebes, A.; Paul, J.; Smit, E.; Smith, E. G.; Sutton, J. K.; Wilcox, M. H.; Dingle, K. E.; Peto, T. E.; Crook, D. W.; Wilson, D. J.; Wyllie, D. H. *Virol. J.* **2013,** *10,* 335.
50. Bull, R. A.; Eden, J. S.; Luciani, F.; McElroy, K.; Rawlinson, W. D.; White, P. A. *J. Virol.* **2012,** *86,* 3219–3229.
51. Cotten, M.; Petrova, V.; Phan, M. V.; Rabaa, M. A.; Watson, S. J.; Ong, S. H.; Kellam, P.; Baker, S. *J. Virol.* **2014,** *88,* 11056–11069.
52. United States National Institute of Allergy and Infectious Diseases (NIAID). *NIAID Biodefense Research Agenda for Category B and C Priority Pathogens;* NIAID: Bethesda, MD (accessed Sep 11, 2012)

53. Debbink, K.; Lindesmith, L.; Baric, R. S. *Clin. Infect. Dis.* **2014,** *58,* 1746–1752.
54. Baehner, F.; Bogaerts, H.; Goodwin, R. *Clin. Microbiol. Infect.* **2016,** *22,* 136–139.
55. Xia, M.; Farkas, T.; Jiang, X. *J. Med. Virol.* **2007,** *79* (1), 74–83.
56. Mason, H. S.; Ball, J. M.; Shi, J. J.; Jiang, X.; Estes, M. K.; Arntzen, C. J. *Proc. Natl. Acad. Sci. USA* **1996,** *93,* 5335–5340.
57. Chachu, K. A.; LoBue, A. D.; Strong, D. W.; Baric, R. S.; Virgin, H. W. *PLoS Pathog.* **2008,** *4,* e100236.
58. Atmar, R. L.; Bernstein, D. I.; Harro, C. D.; Al-Ibrahim, et al. *N. Engl. J. Med.* **2011,** *365,* 2178–2187.
59. Jones, M. K.; Watanabe, M.; Zhu, S.; Graves, C. L.; Keyes, L. R.; Grau, K. R.; Gonzalez-Hernandez, M. B.; Iovine, N. M.; Wobus, C. E.; Vinjé, J.; Tibbetts, S. A.; Wallet, S. M.; Karst, S. M. *Science* **2014,** *3462,* 755–759.
60. Ettayebi, K.; Crawford, S. E.; Murakami, K.; Broughman, J. R.; Karandikar, U.; Tenge, V. R.; Neill, F. H.; Blutt, S. E.; Zeng, X. L.; Qu, L.; Kou, B.; Opekun, A. R.; Burrin, D.; Graham, D. Y.; Ramani, S.; Atmar, R. L.; Estes, M. K. *Science* **2016,** *353,* 1387–1393.
61. Richards, G. P. *Food Environ. Virol.* **2012,** *4,* 6–13.

INDEX

A

Animal models
  emetic activities of SEs, 12–13

B

Bio-plex solution hybridization
  human norovirus
    molecular tests, 216

C

Campylobacter, 56
  advances in research, 79
    developing valid vaccine, 80
    finding relevant and convenient animal models, 80
    studying on pathogenesis, 80–81
  biofilm formation
    genes involved in, 68–69
  biological characteristics, 57–58
  campylobacteriosis, 56
    C. jejumi infection, 57
    symptoms, 57
  diseases
    acute complications, 62–63
    chronic complications, 63–64
    enteritis, 61–62
  epidemiology
    incidence, 59
    infectious dose, 60–61
    infectious routes, 59–60
    population factors, 61
  genomics, 64
    adherence/invasion related genes, 67–68
    biofilm formation, 68
    CPS, 66–67
    flagella, 65
    LOS, 66
    protein secretion systems, 67
    toxins, 65–66
  identification and detection
    biosensors, 78
    direct examination, 73
    immunoassay methods, 77
    MALDI-TOF MS, 77
    membrane filtration, 75
    PCR, 76–77
    traditional culture method using selective agar, 74
    typing, 75–76
  pathogenesis
    adherence, 70
    bacterial survival and immune response, 72–73
    invasion, 71
    mucosal translocation, 71–72
  prevention, 78–79
Capsular polysaccharide (CPS), 66–67
Cytolethal distending toxin (CDT), 65

D

DNA microarrays
  human norovirus
    molecular tests, 216–217
DNA sequencing
  human norovirus
    molecular tests, 217–218

E

Electrical impedometry
  Shigella
    flow cytometry, 189

isolation and identification, 189
LAMP, 190
phage bio-control, 189–190
Enterotoxigenic *Escherichia coli* (ETEC), 124
biological characteristics, 125
diseases, 125–126
epidemiology
outbreak, 126, 128
surveillance, 128–130
genomics
chromosome, 131–132
plasmids, 131
identification and detection, 139–140
outbreaks Worldwide, 127
pathogenesis
bile salts, 136–137
binding and colonizing intestine, 132–133
cell-contact, 138
glucose/*cAMP,* 137
iron, 137–138
LT and ST toxins, 133–135
mechanism(s), 139
oxygen, 138
quorum-sensing, 137
regulation of virulence factor expression, 136
virulence factors, 135–136
prevention
change of lifestyle, 143
food safety, 142–143
vaccination, 141
treatment, 140–141
Escherichia coli, 94
biological characteristics, 95
diseases, 96–97
epidemiology
O157:H7, 98
outbreaks of O157:H7, 99
serotype O157, 100
transmission modes for, 97
genomics, 100–101
identification and detection
enrichment and inoculating on, 107–109
immunological methods, 109–110
immunomagnetic separation, 109
molecular detection methods, 109–112
pathogenesis
LEE, 104
pO157 plasmid, 105–106
potential virulence factors, 106–107
Stx, 102–104
prevention, 113–115
Stx, 94
treatment, 112

H

Human norovirus, 206
biological characteristics, 207
genome, 207
classification
genogroups, 209
genotypes, 209
prototype, 208
epidemiology, 211
food contamination and, 212
person-to-person contact, 212
illness, 210–211
laboratory testing, 212–213
electron microscopy, 213
enzyme immunoassays, 214
molecular tests, 214–218
sample preparation, 213
lessons learned from
cruise ships, 221–228
food service, 220–221
regulatory requirements, 229–232
pathogenesis, 210
prevention
personal hygiene, 219–220

vaccines, 218–219
prevention
disinfection, 220
treatment, 218

I

Immunology-based typing, 39–40
In-frame gene deletion
primers used to generate and verify, 164

L

Lipooligosaccharides (LOS), 66
Listeria monocytogenes, 152
biological characteristics, 153–154
disease, 154
epidemiology, 154–156
genomics, 156–157
approaches for, 159–160
genetic manipulation, 157–159
in-frame gene deletion, 162–165
new suicide plasmid for, 161–162
selection for plasmid excision
with antisense secy, 160–161
identification and detection, 166–170
pathogenesis, 165–166
Locus of enterocyte effacement (LEE), 104

M

Multilocus sequence typing (MLST), 40
Multiplex PCR
Shigella
ATP bioluminescence, 188
isolation and identification, 188

N

National norovirus outbreak surveillance network (CaliciNet)
human norovirus
regulatory requirements, 230–231
National outbreak reporting system (NORS)
human norovirus
regulatory requirements, 230
Norovirus sentinel testing and tracking network (NoroSTAT)
human norovirus
regulatory requirements, 231

O

Outer membrane vesicles (OMVS), 35

P

Pulsed field gel electrophoresis (PFGE), 40

R

Real-time quantitative PCR (RT-QPCR)
human norovirus
molecular tests, 215–216
Reverse transcription-PCR (RT-PCR)
human norovirus
molecular tests, 214–215

S

Salmonella, 26
autophagy
anti-tumor therapies, 42–43
biological characteristics, 27–28
diseases, 30
asymptomatic carrier state, 32
bacteremia and iNTS diseases, 32
enteric fever, 31
gastroenteritis, 31
enteric infections, 26
epidemiology
antibiotic resistance, 30
incidence, 28
infectious dose, 29
infectious routes, 29
peak seasons, 30
susceptible populations, 30
future challenges, 43–44
genomics, 33

adhesion and invasion related genes, 36
biofilm-related genes, 37
function of common proteins of T3SS, 34
OMVS, 35
phases, 36
phoPQ two-component regulatory system, 35
plasmids, 36
polysaccharide Vi capsule, 35
SPIs, 33–35
identification and detection
PCR qPCR, 41
traditional culture methods, 38–39
typing and subtyping, 39–41
pathogenesis, 37–38
prevention, 43
treatment, 41–42
vaccines, 27, 42
Shiga toxin (Stx), 94, 102–104
Shigella, 180
biological characteristics, 181–182
epidemiology, 182
food borne shigellosis, 183–184
molecular, 184
food items associated with outbreaks, 191–196
source of contamination, 192
HACCP for outbreak management, 197
isolation and identification, 184–186
biophysical and biochemical methods, 189–190
detection medium, 186–188
immunological method, 189
pathogenesis, 190
virulence factors, 191
treatment and prevention, 196–197
Staphylococcal enterotoxins (SE)
animal models, 10–11
cell-based approaches for examination
immunological methods, 14, 16
molecular methods, 11, 14
oligonucleotide primers of, 15
cell-based assay, 11
characteristics, 6–10
gene size and encoding gene location of, 8–9
mass spectrometry (MS)-based methods, 16
Staphylococcal food poisoning (SFP), 5
foods associated with, 6
Staphylococcus aureus, 4
colonizes, 5
foodborne pathogens, as, 5

V

Vessel sanitation program
human norovirus
regulatory requirements, 229

W

Whole genome sequencing (WGS), 41
Winter vomiting disease, 206

PGSTL 07/05/2018